# SCIENCE EXPERIMENTS

## BIOLOGY AND ECOLOGY

BY
TAMMY K. WILLIAMS

COPYRIGHT © 1995 by Mark Twain Media, Inc.

ISBN 1-58037-084-5

Printing No. CD–1817

Mark Twain Media, Inc., Publishers
Distributed by Carson-Dellosa Publishing Company, Inc.

T 56252

# CONTENTS

# INTRODUCTION

Both inside and outside of each of us are miraculous processes and cycles that go on without our even paying much attention. Some of these processes involve a single cell, a single nerve ending, or a single organism. Other processes involve entire populations of organisms over many generations. In order to fully understand these processes, certain skills are necessary. These skills include the use of equipment such as balances or microscopes. Weighing accurately, making careful observations, and designing good experiments are also necessary.

In this book these skills are put to use exploring the processes that go on inside living organisms as well as processes that go on around organisms and affect their populations. Units included are laboratory skills, biology, and ecology.

In the laboratory skills section, objects are classified by different appearances, household items are given nonsense names after their characteristics are observed, cheek cells from inside students' mouths are taken and observed, experiments are conducted to measure a water droplet's splatter size and a ball's bounce height, balloon rockets are constructed, and, finally, an edible test of measurement skills is conducted.

In the biology section, the interiors of onion cells are observed, the amount of water in different types of food is calculated, leaf color change is investigated, safe blood transfusions are modelled, crackers are chewed and spit out to test starch digestion, human horsepower is calculated, and microbes are grown to demonstrate the effectiveness of germ killers.

In the ecology section, "birds" that feed on toothpicks have to compete to find enough food to survive; "bears" try to gather enough food to survive; "predators" are released to capture prey that aren't able to band together to defend themselves; "pesticides" are eaten by grasshoppers, shrew, and hawks, destroying their food chain; "deer" population cycles are modelled; and neighborhoods meet to debate issues that affect everyone in the area.

From the microscopic to entire populations of organisms, important processes affect us all. Each process is critical to the survival of the individual as well as the population. Enjoy the activities, and don't be afraid to get your hands dirty.

> "You learn to do what you do and not something else."
> Gerald Unks, Ed 41, UNC-CH, 1984

# LABORATORY SKILLS  INDEX AND MATERIALS LIST

Date: _____ Names: _____

# CLASSIFICATION

INTRODUCTION: How do scientists decide that crocodiles belong in one family of organisms while alligators belong in another family?

OBJECTIVE: In science, organisms are grouped according to different or similar characteristics. This process, called classification, allows for the study of the similarities and differences between organisms. In this activity, we will practice classifying a set of colored and marked cards in order to examine the flexibility that exists when classifying things.

PROCEDURE:
1. A set of marked colored cards are in the envolope provided.
2. Your goal is to design as many classifications systems for those cards as possible.
3. As you decide on a classification system:
    a. Arrange the cards into the system.
    b. In the chart below, identify CATEGORIES and MEMBERS in the spaces provided. Members may be drawn in or described.
    c. Draw vertical lines in the charts to separate categories.
    d. List all the members of a category in the column below the category name.

Example: SYSTEM 1: <u>Classify by shape</u>     SYSTEM 2: _____

| CATEGORIES | square | circle | rectangle |
|---|---|---|---|
| M E M B E R S | A2, B1 | B2, A1 | (blank), B1 |

Date: _____ Names: _____

SYSTEM 3: _____ SYSTEM 4: _____

CATEGORIES

M

E

M

B

E

R

S

SYSTEM 5: _____ SYSTEM 6: _____

CATEGORIES

M

E

M

B

E

R

S

Date: _____ Names: _____

SYSTEM 7:_____  SYSTEM 8: _____

CATEGORIES

M

E

M

B

E

R

S

QUESTIONS:

1. What characteristics about the cards did you use in order to classify them?

_____

_____

2. Does each classification system contain the same categories? Why/why not?

_____

_____

3. Do all categories contain the same members? Why/why not?_____

_____

4. Do you think that each group in the class thought of the same classification systems that you

did? _____

5. Do you think that each group in the class would agree on the same system as THE BEST

SYSTEM? Why/why not? _____

_____

6. Why then is it important for scientists to agree on a single classification system for each particular

group of items/organisms? _____

_____

_____

Date: _____ Names: _____

# DICHOTOMOUS KEY

INTRODUCTION: Once plants and animals have been assigned by scientists to certain families, how do you figure out their names or species? This is done by using a device called an identification key.

OBJECTIVE: In science, organisms are identified and classified according to characteristics that they possess. These characteristics may be either similar to or different from those of other organisms. When differences are observed so that the presence or absence of a characteristic determines which category the organism (or object) falls into, the identification tool is called a DICHOTOMOUS KEY. In this activity, we will use a dichotomous key to give household items nonsense names.

PROCEDURE: 1. For each item provided, start with description number 1 and follow the instructions at the end of the line of the description that fits your item until the end of the line provides a name for that item.
2. In the space beside each nonsense name provided, write in the actual name of the household item.

1a. Object is partly or completely made of metal ........................... go to 2
1b. Object has no metal on it ...................................................... go to 16

2a. Object has nonmetal parts ................................................... go to 3
2b. Object is completely made of metal ....................................... go to 5

3a. Object is less than 10 cm in length ........................... whippersnapper _____
3b. Object is 10 cm or greater in length ........................................ go to 4

4a. Object is pointed at one end ...................................... tapered doodad _____
4b. Object is not pointed at one end ............................... common doodad _____

5a. Object is greater than 10 cm ................................................... go to 6
5b. Object is 10 cm or less ........................................................... go to 9

6a. Object has a twisted area ................................................. thingamajig _____
6b. Object has no twisted area ...................................................... go to 7

7a. Object has three or more prongs ............. left-handed monkey wrench _____
7b. Object has no prongs ............................................................. go to 8

Date: _____ Names: _____

8a. Object has a cutting edge ................................................ geegaw _____

8b. Object has no cutting edge .......................................... scooperdoo _____

9a. Object has spiral grooves ............................................ go to 10

9b. Object has no spiral grooves ....................................... go to 11

10a. Object has a hole ........................................................ cashew _____

10b. Object has no hole ...................................................... whatsit _____

11a. Outside edge is a circle ............................................. go to 12

11b. Outside edge is not a circle ....................................... go to 13

12a. Object is silver-colored ............................................. quinto _____

12b. Object is not silver-colored ........................................ uno _____

13a. Object is silver-colored ............................................. go to 14
_____

13b. Object is not silver-colored ........................................ go to 15

14a. Object is less than 4 cm in length .............................. micro whatnot _____

14b. Object is 4 cm or more in length ............................... macro whatnot _____

15a. Object is brass-colored ............................................. skyhook _____

15b. Object is not brass-colored ....................................... dingus _____

16a. Object is white ........................................................... go to 17

16b. Object is not white ..................................................... go to 24

17a. Object has holes ........................................................ wadget _____

17b. Object has no holes .................................................... go to 18

18a. Object is a circle in at least one dimension .............. go to 19

18b. Object is not a circle in any dimension ...................... go to 20

Date: _____ Names: _____

19a. The circumference of the circular dimension is 6 cm or less... bric-a-brac _____

19b. The circumference of the circular dimension is greater than 6 cm ...........

.......................................................................... roundabout _____

20a. Object is made of plastic ............................................................. go to 21

20b. Object is not made of plastic ...................................................... go to 23

21a. Object has 3 or more prongs ................................................. doohickey _____

21b. Object has no prongs ............................................................. go to 22

22a. Object has a cutting edge ....................................................... gismo _____

22b. Object does not have a cutting edge ....................................... flim flam _____

23a. Object appears to have a string running through its center ........ wickey _____

23b. Object does not appear to have a string running through its center .......

.......................................................................... scrubadub _____

24a. Object is made of plastic ......................................................... go to 25

24b. Object is not made of plastic .................................................. go to 28

25a. Outer edge of the object is round ......................................... go to 26

25b. Outer edge of the object is not round ......................... whatchamacallit _____

26a. Object has holes .................................................................... go to 27

26b. Object has no holes ............................................................. spinaroo _____

27a. Object has 2 holes ................................................................. bihole _____

27b. Object has 4 holes ............................................................... tetrahole _____

28a. Object is made of glass ......................................................... seethru _____

28b. Object is not made of glass .................................................. go to 29

29a. Object is yellow in color ....................................................... screecher _____

29b. Object is not yellow in color .................................................. soaky _____

Date: _____ Names: _____

# METRIC MEASUREMENT (LENGTH)

INTRODUCTION: If your hand is 3 inches wide, how many centimeters wide is it? Which metric unit is closest to the length of 1 yard?

OBJECTIVE: In this activity, we will review metric units for measuring distance or length—the meter, decimeter, centimeter, and millimeter. We will also use these units to estimate and then measure the sizes of various objects around the room.

Meterstick: 1 meter (m) = 10 decimeters (dm) OR 100 centimeters (cm) OR 1,000 millimeters (mm)
Here is a visual representation of a meterstick.

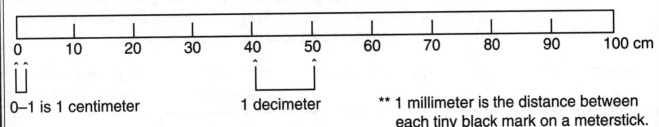

0–1 is 1 centimeter          1 decimeter          ** 1 millimeter is the distance between each tiny black mark on a meterstick.

PROCEDURE:
1. Use a meterstick to measure the objects listed in the chart below. Make sure you use the metric side of the meterstick (with numbers to 100 cm, not 36 inches).
2. Measure the objects in the units listed. Write the unit abbreviation after the measurement you get (example: instead of 47.5, write 47.5 cm).

| OBJECT | MEASUREMENT | UNITS |
|---|---|---|
| Length of your table | | Meters (m) |
| Width of your table | | Decimeters (dm) |
| Length of a piece of paper | | Centimeters (cm) |
| Width/thickness of a pencil | | Millimeters (mm) |

3. Which unit above is closest to the following size:
   a. the thickness of a fingernail? _____
   b. the width of a finger? _____
   c. the width of a hand? _____
   d. longer than your leg? _____

Date: _____ Names: _____

4. Keep the sizes of each of the metric units in mind. For each object listed in the chart below:

    a. Choose the most appropriate unit of measurement (m, dm, cm, mm) and record that unit in the chart in the "Unit Chosen" column.

    b. Estimate the size of that object using the units you choose and the "body parts" in number 3a–d above. You may actually lay fingers side-by-side along an object to see how many centimeters long it is. Record your estimates in the chart below under the "Estimate" column.

    c. Get up and measure the objects listed using the units that you chose. Record your measurements in the chart below the "Measurement" column. You do not have to measure the items in the order listed.

| OBJECT | UNIT CHOSEN | ESTIMATE (WITH UNITS) | MEASUREMENT (WITH UNITS) |
|---|---|---|---|
| Height of table | | | |
| Length of tabletop | | | |
| Height of classroom door | | | |
| Thickness of tabletop | | | |
| Width of cabinets | | | |
| Thickness of a pencil lead | | | |
| Width of your table leg | | | |

QUESTIONS:

1. Which unit might be best used to measure:   a. shoe length? _____

       b. thickness of hair strands? _____

       c. a bus length? _____

       d. width of a door? _____

       e. length of a hallway? _____

       f. height of the letter "E"? _____

       g. length of a pencil?_____

2. How is the metric system simpler to use than English units (like inches, feet, and yards)?

_____

_____

Date: _____   Names: _____

# METRIC MEASUREMENT (VOLUME)

INTRODUCTION: The volume of a cube can be calculated by multiplying its length times its width times its height. How could you figure out the volume of a rock that has broken and chipped edges? How could you figure out the volume of a bag of marbles without doing a lot of math?

OBJECTIVE: In this activity, we will learn how to read the volume of a liquid in a graduated cylinder measuring milliliters (mL) by reading the meniscus of the liquid (see diagram below). When most liquids are placed in tall, narrow containers, they creep up the walls of the container a little due to capillary action. This results in the surface of the liquid appearing to be curved. The bottom of this curve is known as the MENISCUS, and best represents the actual volume of liquid in the cylinder. We will also learn how to measure the volume of odd-shaped objects.

Graduated cylinder:

60 mL

← meniscus

PROCEDURE: 1. Pour the colored liquid from the beaker at your lab station into the graduated cylinder.

2. Sit the graduated cylinder flat on the counter top.

3. Bend over so that the water level is at eye level and look for the meniscus.

4. Record the number of milliliters of liquid (to the nearest one-half mL) in the chart on the next page. This step will be done before each object is lowered into the liquid. Since this prepares us to measure the first object, record the liquid volume in the first box under "Beginning Volume" (second column).

5. Once a starting liquid volume has been measured, gently lower an object into the liquid. The amount that the water rises (amount of water displaced) is equal to the volume of the object.

6. Read the new volume at the meniscus and record it in the chart under "Volume of Liquid & Object" for that object (first column).

7. To calculate the volume of the object alone, subtract the "Beginning Volume" from the "Volume of Liquid & Object" (column 2 from column 1).

8. Repeat the above steps for each of the remaining objects.

Date: _____ Names: _____

| OBJECT | VOLUME OF LIQUID & OBJECT | – | BEGINNING VOLUME (LIQUID) | = | VOLUME OF OBJECT |
|--------|--------------------------|---|---------------------------|---|------------------|
| Nail | | | | | |
| Screw | | | | | |
| Penny | | | | | |
| Rock | | | | | |

QUESTIONS:

1. Why is it necessary to recheck the starting volume of liquid before each object is added?

_____

_____

_____

2. What kind of error would result if you read the liquid volume where the liquid touches the wall of the cylinder rather than at the meniscus?

_____

_____

_____

3. How does this "measuring volume by difference" method compare with measuring volume using math for these odd-shaped objects?

_____

_____

_____

_____

_____

_____

Date: _____ Names: _____

# METRIC MEASUREMENT (MASS/WEIGHT)

INTRODUCTION: How can you figure out how much of your pencil gets "eaten" by a pencil sharpener each time you sharpen a pencil? How can you figure out how much a gulp of water is?

OBJECTIVE: In this activity, we will become familiar with the parts of a triple-beam balance that is used to measure mass, and we will practice measuring the mass of different objects. Following this, we will learn how to "weigh-by-difference" to find the mass of different objects.

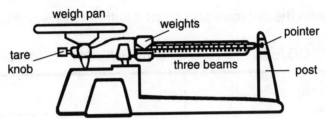

PROCEDURE: To "ZERO A BALANCE"

1. Check to make sure that the balance is clean. Wipe and clean it if necessary.

2. Move all weights to the left of the balance (next to the weigh pan).

3. Look to see if the pointer line is perfectly in line with the mark on the post. This indicates whether the balance is zeroed.

4. If the lines do not meet, adjust the "TARE KNOB," which is located underneath the weigh pan, by turning it a little and observing its effect. You should be able to zero the balance by repeating this procedure.
** If you cannot zero the balance, ASK FOR ASSISTANCE!

To Weigh Objects

1. Use the following steps to weigh each object listed in the chart (in grams) and, record its weight in the chart.

2. Make sure the balance is zeroed and the weigh pan is clean.

3. Place an object on the weigh pan.

4. Move the weights on the beams until the pointer balances at the white mark on the post. To do this, move the small weight to the right. If it is too light to balance the object, move it back to the left (to 0) and try the next larger weight. Continue this until one of the weights can be placed so that the pointer is both above and below the post line.
** Make sure that the two larger weights fall into notches as you move them on the beams.

Date: _____ Names: _____

5. Weights can be measured as accurately as the nearest tenth of a gram by positioning the smallest weight.

6. Once the weights have been positioned so that the beam pointer aligns with the mark on the post, add each of the marked weights together to get a total. Remember the smallest weight marks single grams, and the lines between the numbers on that beam mark tenths of grams. The medium-sized weight marks tens of grams, and the largest weight marks hundreds of grams.

7. Record the total mass in the chart below under "Weight." Write in units.

8. Clean the balance and store it with all the weights on zero.

| OBJECT | WEIGHT (IN GRAMS) |
|---|---|
| Small paper clip | |
| Two small paper clips | |
| Large paper clip | |
| Two large paper clips | |
| One penny | |
| Empty beaker | |
| Something you choose: _____ | |

<u>To Weigh By Difference</u>
1. When doing this, we will be weighing an object, removing from or adding to the object, and then reweighing the object to see how much was taken away or added.

2. First, weigh each of the objects listed in the chart on page 15. Record their weights under the column "Weight Before."

3. For each object remove from or add to it by:
   a. putting the sponge in water.
   b. drinking a swallow of water from the cup.
   c. sharpening the pencil.

4. Reweigh each item after step #3 and record its new weight under "Weight After."

Date: _____  Names: _____

5. To find the amount of change (weight gained or lost) subtract the smaller number from the larger number. If the starting number is larger, the weight was lost. If the ending weight was larger, then the weight was gained.

6. Record the units used as well as whether weight was gained or lost.

7. Store the balance clean and dry and with all weights on zero.

| OBJECT | WEIGHT BEFORE | WEIGHT AFTER | WEIGHT CHANGE (GRAMS) |
|---|---|---|---|
| Dry sponge (put in water) | | | |
| Cup of water (take a sip) | | | |
| Pencil (sharpen) | | | |

QUESTIONS:

1. Why is it important to make sure that the weigh pan is clean before weighing objects?

_____

_____

_____

2. How does "weighing by difference" compare to something like saving pencil shavings and weighing them to find out how much was sharpened off? _____

_____

_____

_____

_____

Date: _____ Names: _____

# MICROSCOPE USE
# OBSERVING HUMAN CHEEK CELLS

INTRODUCTION: What do the cells in your mouth look like? What is inside of them?

OBJECTIVE: In this activity, we will practice proper use of the microscope in order to observe human cheek cells.

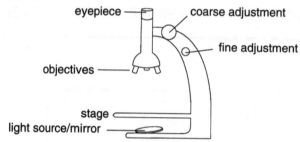

PROCEDURE:

1. Clean the microscope slide by wiping it with lens paper/tissue.

2. With the dropper provided, place a drop of water onto the slide in the center.

3. Using a toothpick, gently scrape the inside wall of your cheek (in your mouth) with one end of the toothpick.

4. Gently stir that end of the toothpick in the water droplet on the microscope slide. Be careful not to smear the water droplet.

5. Squeeze one drop of iodine from the dropping bottle labeled "IODINE" into the water/cell droplet on the microscope slide.

6. Add a small glass coverslip to the droplet by standing the coverslip on its edge beside the droplet and allowing it to gently fall over onto the droplet. This will squish air bubbles from underneath the coverslip as it falls.

7. Place the slide onto the microscope stage and clip it into place using the stage clips.

8. You might need to adjust the mirror of the microscope or move the microscope so that light will reflect up through the stage and slide. If you carry the microscope, use two hands—one underneath and the other on its frame.

9. Begin focusing on the lowest power with the objective rolled down as far as it will go.

10. Once you have focused on low power, without moving the adjustment knob, switch objectives to medium power. Refocus using the fine adjustment knob.

16

Date: _____ Names: _____

11. Repeat step #10 for high power, except when turning the objective watch from the side to make sure that you do not crack the slide and coverslip.

12. ON THE HIGHEST POSSIBLE POWER YOU CAN FOCUS, sketch the cells that are visible in the "field of view" (circular area visible through eyepiece) as precisely as possible in the circle below.
** If a something that looks like a needle is visible through your scope, ignore it. It is a pointer that you can use to show other people an exact structure on the slide.

13. When your sketch is complete, wash your slide and coverslip, throw away your used toothpick, clean microscope stage, and store the scope on LOW POWER with the stage rolled down as far as it will go.

Cell sketch:

QUESTIONS:

1. What job does the drop of water seem to do? _____
_____

2. What does the iodine do to the cheek cells (and everything else it touches)? _____
_____
_____

3. Why do you need to watch from the side of the microscope when switching to HIGH POWER?
_____
_____

4. What do your cheek cells look like? _____
_____
_____

Date: _____ Names: _____

# MAKING A HYPOTHESIS

INTRODUCTION: Do you know what will happen if you mix vinegar and baking soda? Will the temperature change? Will bubbles form? Guessing what will happen is called making a hypothesis.

OBJECTIVE: A hypothesis is a "best guess" because the outcome of a question is guessed using only what is known before the question is tested. The hypothesis is then tested using an experiment. During an experiment, data or information is collected to check the accuracy of the hypothesis. Finally, using the results of the experiment, the hypothesis may be supported as correct or it may be changed. In this activity, we will make hypotheses about how combining liquids with two different chemicals will affect the temperature of each liquid.

PROCEDURE: <u>CHEMICAL A AND WATER</u>

1. Place your hypothesis for "Chemical A and Water" below. What do you think will happen?

2. Measure 5 mL of water in the graduated cylinder and pour it into a test tube.
3. Place the test tube in the test tube rack and gently place the thermometer in the water in the test tube.
4. After 1 minute, record the temperature of the water and record this temperature in the chart on page 18 under "Start."
5. Using the metal spatula, add 3 pellets of "Chemical A" to your test tube. DO NOT TOUCH THE PELLETS WITH YOUR SKIN!
6. Observe and record the temperature of the water every 15 seconds for 3 minutes.
7. At the end of 3 minutes, remove the thermometer, pour the contents of the test tube down the drain, and rinse the test tube and thermometer.
8. Graph the data from the chart onto the graph sheet on the next page. Use a SOLID LINE to connect the dots of data.

<u>CHEMICAL B AND VINEGAR</u>

9. Place your hypothesis for "Chemical B and Vinegar" below. What do you think will happen?

10. Repeat the exact procedure used for Chemical A, EXCEPT:
    a. In #2 above, begin with 5 mL of vinegar.
    b. In #5 above, add 1/2 teaspoon of "Chemical B."
11. Graph the data for Chemical B onto the graph on the next page. Use a BROKEN LINE to connect the data points.

Date: _____  Names: _____

DATA:

## TEMPERATURE CHANGE (DEGREES CELSIUS)

| | Start | 15 | 30 | 45 | 60 | 75 | 90 | 105 | 120 | 135 | 150 | 165 | 180 |
|---|---|---|---|---|---|---|---|---|---|---|---|---|---|
| CHEMICAL A | | | | | | | | | | | | | |
| CHEMICAL B | | | | | | | | | | | | | |

Time (seconds)

## TEMPERATURE VERSUS TIME

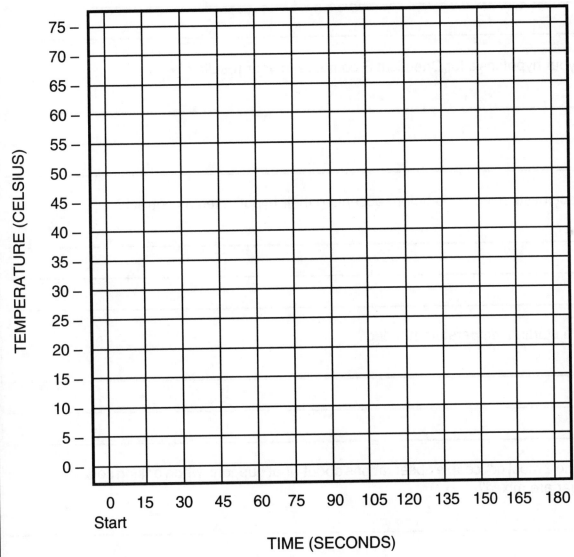

TIME (SECONDS)

Date: _____ Names: _____

QUESTIONS:

1. What happened to the water temperature when Chemical A was added?

_____

2. What happened to the vinegar temperature when Chemical B was added?

_____

3. How did your hypothesis for Chemical A compare to your results?

_____

_____

_____

4. How did your hypothesis for Chemical B compare to your results?

_____

_____

_____

5. What do scientists need to do before accepting their hypotheses as correct?

_____

_____

_____

6. Why was a starting temperature needed?

_____

_____

_____

7. What are the 2 most noticeable observations about what happens when baking soda (Chemical B) and vinegar are mixed?

_____

_____

_____

Date: _____    Names: _____

# SCIENTIFIC METHOD
# SHAPE OF WATER SPLATTER VERSUS DROP HEIGHT

INTRODUCTION: If you dribble Kool-Aid while you are standing when you pour it, will it splatter more or less than if you are sitting when you pour it?

OBJECTIVE: In this activity, we will practice using the scientific method while investigating the effect of drop height on the size and shape of water droplet splatters when they land. We will be careful to change only the one item whose effect we will observe. This is called the EXPERIMENTAL VARIABLE. All of the other conditions must be kept completely identical. These conditions are called CONTROLS.

PROCEDURE:    1. As you follow the instructions to complete the water droplet investigation, fill in the steps of the scientific method by writing in what you do at each step of the investigation where it matches a step in the scientific method.

STATE PROBLEM: _____

_____

GATHER INFORMATION (name sources of information):

a. _____    b. _____

c. _____    d. _____

MAKE HYPOTHESIS: _____

_____

EXPERIMENT: _____

_____

_____

RECORD DATA (list examples of data): a. _____

b. _____    c. _____

FORM CONCLUSION: _____

_____

Date: _____ Names: _____

2. Add 2 drops of food coloring to your beaker.

3. Add 100 mL of water and mix.

4. Partially fill the glass dropper with colored water.

5. Measure the heights listed in the chart using a meterstick positioned with one end on the splatter paper and the other end measuring dropper height.

6. From each height, drop 3 drops of water (in different places on the paper).

7. Measure the diameter of the splatter in MILLIMETERS, and record each trial size in the chart. Add the sizes to get a total, and divide by 3 to find the average size of each splatter.

8. Repeat this process for each height.

9. For each drop height, write a description of the splatter in the chart, also. Examples: "Drop is very round," "Drop broke apart," or "Drop is surrounded by little splatters."

10. On the graph, plot the average splatter size versus drop height.

DATA:

### DIAMETER OF DROP SPLATTERS (mm)

| Drop Height | Trial 1 | Trial 2 | Trial 3 | Total | Average | Description |
|---|---|---|---|---|---|---|
| 5 cm | | | | | | |
| 10 cm | | | | | | |
| 20 cm | | | | | | |
| 40 cm | | | | | | |
| 80 cm | | | | | | |

Date: _____    Names: _____

## DROP HEIGHT VERSUS SPLATTER SIZE

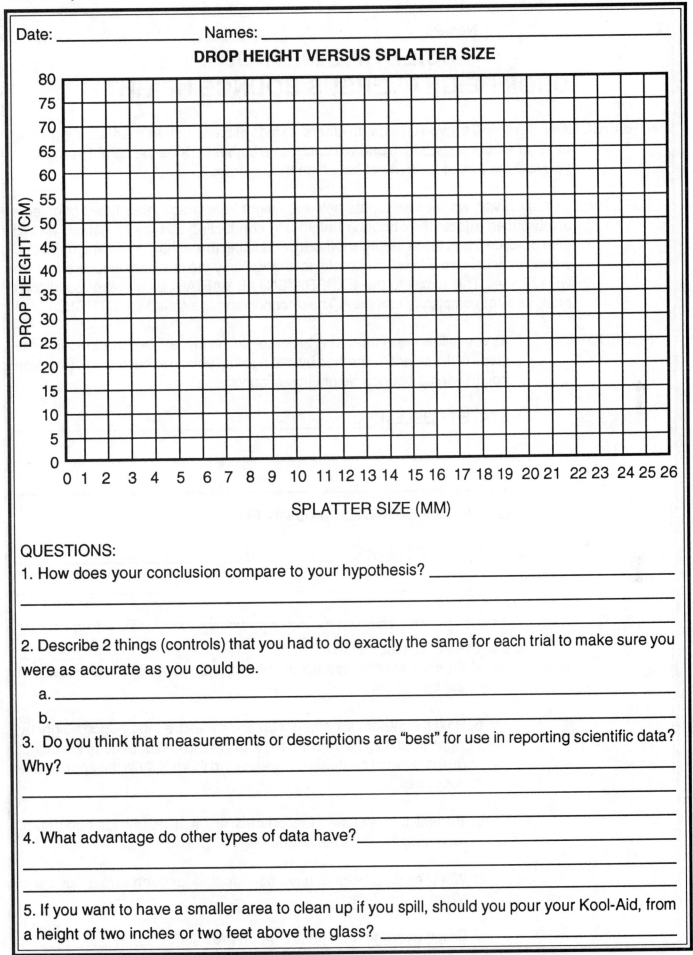

SPLATTER SIZE (MM)

QUESTIONS:

1. How does your conclusion compare to your hypothesis? _____
_____
_____

2. Describe 2 things (controls) that you had to do exactly the same for each trial to make sure you were as accurate as you could be.

   a. _____

   b. _____

3. Do you think that measurements or descriptions are "best" for use in reporting scientific data? Why? _____
_____
_____

4. What advantage do other types of data have?_____
_____
_____

5. If you want to have a smaller area to clean up if you spill, should you pour your Kool-Aid, from a height of two inches or two feet above the glass? _____

23

Date: _____ Names: _____

# SCIENTIFIC METHOD
# DROP HEIGHT VERSUS BOUNCE HEIGHT

INTRODUCTION: If you spill a toybox filled with different kinds of balls such as marbles, ping-pong balls, or rubber balls, which one will bounce up highest? Does the height from which you spill the box of balls affect how high they bounce?

OBJECTIVE: In this activity, we will investigate how the height from which three types of objects are dropped affects their bounce heights. In conducting this investigation, we will also practice using the scientific method: conducting three trials, graphing data, and writing conclusions. We will be careful to change only the one item whose effect we will observe. This is called the EXPERIMENTAL VARIABLE. All other conditions must be kept completely identical. These conditions are called CONTROLS.

PROCEDURE:

1. As you follow the instructions to complete the investigation below, fill in the steps of the scientific method by writing what you do at each step in the drop height versus bounce height investigation.

2. STATE PROBLEM: _____

_____

_____

GATHER INFORMATION (skip this item)

3. STATE HYPOTHESIS: _____

_____

_____

4. EXPERIMENT (see below):

   a. From each of the identified heights listed in the chart, drop each of the three items.

   b. Hold the meterstick straight up and measure the bounce height by sight. Measure the maximum height of the first bounce of the object (to the nearest centimeter), and record this height in the appropriate data chart.

   c. Repeat this test two more times at each height for each item, recording the data each time.

   d. When all drop heights have been tested for each object, total and average the data.

   e. Graph average data on the graph provided.

Date: _____ Names: _____

## 5. RECORD DATA:

Item 1: _____

| Drop Height | Bounce Height | | | | |
|---|---|---|---|---|---|
| | Trial 1 | Trial 2 | Trial 3 | Total | Average |
| 10 cm | | | | | |
| 20 cm | | | | | |
| 30 cm | | | | | |
| 40 cm | | | | | |
| 50 cm | | | | | |
| 60 cm | | | | | |
| 70 cm | | | | | |
| 80 cm | | | | | |
| 90 cm | | | | | |
| 100 cm | | | | | |

Item 2: _____

| Drop Height | Bounce Height | | | | |
|---|---|---|---|---|---|
| | Trial 1 | Trial 2 | Trial 3 | Total | Average |
| 10 cm | | | | | |
| 20 cm | | | | | |
| 30 cm | | | | | |
| 40 cm | | | | | |
| 50 cm | | | | | |
| 60 cm | | | | | |
| 70 cm | | | | | |
| 80 cm | | | | | |
| 90 cm | | | | | |
| 100 cm | | | | | |

Item 3: _____

| Drop Height | Bounce Height | | | | |
|---|---|---|---|---|---|
| | Trial 1 | Trial 2 | Trial 3 | Total | Average |
| 10 cm | | | | | |
| 20 cm | | | | | |
| 30 cm | | | | | |
| 40 cm | | | | | |
| 50 cm | | | | | |
| 60 cm | | | | | |
| 70 cm | | | | | |
| 80 cm | | | | | |
| 90 cm | | | | | |
| 100 cm | | | | | |

Date: _____ Names: _____

6. FORM CONCLUSION: _____

_____

7. CREATE A LINE GRAPH showing the average data for each of the three objects.

**AVERAGE DROP HEIGHT VERSUS BOUNCE HEIGHT**

DROP HEIGHT (CM)

80 75 70 65 60 55 50 45 40 35 30 25 20 15 10 5 0

0 1 2 3 4 5 6 7 8 9 10 11 12 13 14 15 16 17 18 19 20 21 22 23 24 25 26

BOUNCE HEIGHT (CM)

QUESTIONS:

1. What is the experimental variable for this experiment?_____

2. What are 3 controls for this experiment? a. _____

                                                        b. _____

                                                        c. _____

3. Which of the balls that you tested was most likely to:

    a. bounce low and stay near where it was dropped?_____

    b. bounce high and bounce away from where it was dropped?_____

Date: _____    Names: _____

# SCIENTIFIC METHOD
# ROCKET ENGINES AND NEWTON'S THIRD LAW

INTRODUCTION: What would happen if you stood still while wearing skates and threw baseballs away from you? Would you move? How about if you sat on a skateboard and did the same thing?

OBJECTIVE: Newton's Third Law says that for every action there is an equal and opposite reaction. This means that if you push someone away from you, you are pushed backwards with the same force that you exerted. In the same manner, if a rocket's exhaust exerts 150 Newtons of force against a concrete launch pad, the same force is exerted on the rocket. This is what causes the rocket to be lifted into the air. In this activity, we will investigate this reaction using a balloon as the engine and a straw as the rocket. We will be careful to change only the one item whose effect we will observe. This is called the experimental variable. All other conditions must be kept completely identical. These conditions are called controls.

PROCEDURE:
    A. STATE PROBLEM: _____

_____

    B. GATHER INFORMATION (skip this section)

    C. FORM HYPOTHESIS: _____

_____

    D. EXPERIMENT (see procedure below):
        1. Pull your flight string from where it is attached across the room. String should be 4–6 meters long.
        2. Inflate your balloon to the 3 cm diameter indicated in the chart.
        3. Hold the balloon closed or close it in some other way SO THAT IT CAN BE QUICKLY REOPENED LATER.
        4. Attach the balloon to the straw, OPEN END TOWARDS YOU, CLOSED END TOWARDS THE BOARD. (See diagram.)
        5. Facing the string attachment, hold the flight string level near its free end.
        6. Start the straw at the mark on the string.
        7. Release the rocket (balloon) and allow it to fly down the string.
        8. Holding the string level, measure the distance travelled by the rocket.
        9. Repeat steps 1–8 twice for this diameter, and record the data in the chart.
      10. Repeat the entire procedure for each diameter in the chart.
      11. Graph average distances versus balloon diameter on a computer. Attach your graph print out to your lab sheet. If you don't have access to a computer, use the graph provided on page 28.

Date: _____ Names: _____

## E. RECORD DATA:

**Diameter of Balloon**          **Distance Travelled by Balloon**

| | Trial 1 | Trial 2 | Trial 3 | Total | Average |
|---|---|---|---|---|---|
| 3 cm | | | | | |
| 6 cm | | | | | |
| 9 cm | | | | | |
| 12 cm | | | | | |
| 15 cm | | | | | |

F. FORM CONCLUSION: _____

_____

_____

QUESTIONS:

1. What is the experimental variable in this experiment? _____

2. Name 3 controls for this experiment:  a. _____

b. _____      c. _____

3. How would Newton's Third Law affect you if you threw baseballs away from you while standing on skates or sitting on a skateboard? _____

_____

_____

## BALLOON DIAMETER VERSUS FLIGHT DISTANCE

FLIGHT DISTANCE (M): 7.5, 7.0, 6.5, 6.0, 5.5, 5.0, 4.5, 4.0, 3.5, 3.0, 2.5, 2.0, 1.5, 1.0, .5, 0

BALLOON DIAMETER (CM): 0, 3, 6, 9, 12, 15

Date: _____ Names: _____

# LAB TECHNIQUES GOOD ENOUGH TO EAT

INTRODUCTION: How is cooking similar to chemistry? Both involve following directions well enough to avoid catastrophe!

OBJECTIVE: In this activity, we will practice measuring length, mass, and volume using metric units. The end product will be an indicator of your ability to measure accurately. If your measurement skills are very good, your result will be good enough to eat!

PROCEDURE:

1. Measure 16 cm from the bottom of the large plastic bag and use the pen at your lab station to mark the height.

2. Fill the bag to the line that you marked with ice (from the cooler).

3. MEASURE out 100 mL of ROCK SALT, and pour it over the ice in the large bag.

4. Take a new, small plastic bag and mix the following ingredients into this bag:
   a. 100 mL MILK
   b. 20 mL SUGAR
   c. 2 drops of VANILLA EXTRACT

5. Seal the contents of the small bag (make sure the bag is "ziplocked").

6. Place the small bag inside the large bag of ice.

7. Seal the large bag (with the small bag inside). Make sure the bag is "ziplocked."

8. Move the ice in the large bag around so that it surrounds/covers/supports the small bag.

9. Time 1-minute intervals. At the end of each interval, turn the bag over onto the other side. (Arrange ice cubes so that they surround the small bag inside.)

10. Repeat the 1-minute "flippings" a total of ten times.

11. At the end of the tenth minute (10 1-minute flips), time 30-second intervals and flip the bag every 30 seconds for five minutes (10 30-second flips).

12. At the end of the 20 flips (10 flips 1 minute apart and 10 flips 30 seconds apart), ask your instructor to check to see if your solution has completed its reaction.

Date: _____ Names: _____

13. A. If your solution is approved, you may eat your solution . . . IF YOU DARE!
    B. If your solution is not approved, continue flipping every 30 seconds for an additional 5 minutes. Then you may complete step #13A.

**THIS MAY ALSO BE DONE WITH A BABY FOOD JAR INSIDE A COFFEE CAN WITH A PLASTIC RECLOSEABLE LID. IN THIS SET-UP, ALTERNATE HANDFULS OF ICE AND SALT AND THEN ROLL CAN UNDER FOOT.

**Flip Check-list (X out numbers as you flip)**

| | | | | | | | | | | |
|---|---|---|---|---|---|---|---|---|---|---|
| 1-MINUTE FLIPS | 1 | 2 | 3 | 4 | 5 | 6 | 7 | 8 | 9 | 10 |
| 30-SECOND FLIPS | 1 | 2 | 3 | 4 | 5 | 6 | 7 | 8 | 9 | 10 |

QUESTIONS:

1. What solution do you end up with? _____

2. Instead of an explosion, what might result if you measure inaccurately?

_____

_____

_____

3. Besides successful science experiments, list two other areas where good measurement skills are important.

   a. _____

   b. _____

# LABORATORY SKILLS ANSWER KEYS

## CLASSIFICATION (page 5)

1. Answers will vary, but will include color, size, print color, letters or numbers on them, and so on.
2. No. The system determines what categories will be necessary.
3. No. Category members are determined by the system used.
4. Not exactly, but there will be a lot of overlap because some systems are easy to figure out.
5. No. Different people will prefer different characteristics.
6. Scientists should agree on a best system because it lessens confusion when trying to organize data.

## DICHOTOMOUS KEY (pages 6–8)

3a. wooden snappy clothespin
4a. sharp pencil
4b. unsharpened pencil
6a. wire hanger
7a. metal fork
8a. metal knife
8b. metal spoon
10a. metal nut
10b. bolt
12a. nickel
12b. penny
14a. small paper clip
14b. large paper clip
15a. brass fastener
15b. black bobby pin
17a. white button
19a. white chalk
19b. microwave plate
21a. white plastic fork
22a. white plastic knife
22b. white plastic spoon
23a. white candle
23b. white soap
25b. colored plastic hanger
26b. colored milk lid
27a. colored 2-hole button
27b. colored 4-hole button
28a. glass jar
29a. yellow chalk
29b. colored soap

## METRIC MEASUREMENT (LENGTH) (pages 9–10)

3a. millimeter
  b. centimeter
  c. decimeter
  d. meter

1. Some range of error should be allowed for answers given.
  a. dm/cm
  b. mm
  c. m
  d. dm
  e. m
  f. mm
  g. cm/dm

2. The metric system is based on units of 10 so you can change from one unit to another simply by dividing or multiplying by 10. In the English system, you have to use 3, 12, 36, and so on.

## METRIC MEASUREMENT (VOLUME) (page 12)

1. Water is removed with the object in each trial so the starting volume decreases and may result in a larger end volume being recorded if a new starting volume is not recorded.
2. The volume would be read higher than it actually is.
3. "Measuring by difference" is a lot easier than trying to calculate the volume using math.

## METRIC MEASUREMENT (MASS/WEIGHT) (page 15)

1. A dirty weigh pan means you are weighing dirt and the starting weight is read higher than it should be.
2. "Weighing by difference" is easier than saving shavings (which blow away and stick to stuff).

## MICROSCOPE USE (page 17)

1. The water rinses the cells off of the toothpick.
2. The iodine stains the cells brown so that they contrast with the background and are visible.
3. Watching from the side might prevent you from cracking the slide and coverslip when you switch to high power, which has the longest objective.
4. Answers will vary, but will say something like: "They look like corn flakes with dark spots in the center."

## MAKING A HYPOTHESIS (page 20)

1. The temperature increased.
2. The temperature decreased.
3. Will be determined by hypothesis. Hypothesis will be supported or will not be supported by data.
4. Same as #3.
5. Scientists must test their hypotheses.
6. Without a starting temperature, you won't know if the temperature increases or decreases.
7. Answers may vary, but will probably include: "The solution bubbles over, and the temperature drops."

## SCIENTIFIC METHOD: SHAPE OF WATER SPLATTER VERSUS DROP HEIGHT (page 23)
1. Answer will be determined by hypothesis. Data will either support or not support the hypothesis.
2a. meter stick had to be held exactly straight up;
   b. you had to measure from same point on dropper each time; other controls are acceptable.
3. Measurements are probably best because they can be replicated exactly by different scientists.
4. Other data, like descriptions, give information (like shape) that measurements do not.
5. Pour Kool-Aid from two inches.

## SCIENTIFIC METHOD: DROP HEIGHT VERSUS BOUNCE HEIGHT (page 26)
1. Drop height
2a. meter stick was held perfectly upright;
   b. measurement was made from bottom of each ball;
   c. drop was made onto same surface each time
3a. lowest bouncing ball (determined by items used and data recorded)
   b. highest bouncing ball (determined by items used and data recorded)

## SCIENTIFIC METHOD: ROCKET ENGINES AND NEWTON'S THIRD LAW (page 28)
1. Balloon diameter
2a. measuring from same end of straw each time;
   b. holding string perfectly level each time;
   c. not allowing rocket to move until measurement is complete each time
3. When you have no friction holding you in place, throwing objects one direction will push you in the opposite direction.

## LAB TECHNIQUES GOOD ENOUGH TO EAT (page 30)
1. Ice cream
2. Bad tasting ice cream would result.
3a. following a recipe when cooking;
   b. cutting out pieces of a pattern/puzzle; other items may be accepted.

# BIOLOGY INDEX AND MATERIALS LIST

Date: _____ Names: _____

# MITOSIS/MEIOSIS

INTRODUCTION: What happens inside cells to help them produce new cells? If you could watch a single cell divide, what would you see?

OBJECTIVE: Organisms grow and repair themselves by reproducing cells. To reproduce a cell, everything inside the cell must be duplicated: cytoplasm, mitochondria, endoplasmic reticula, vacuoles, and the cell membrane, as well as the nucleus with the DNA inside. During cell division, the DNA breaks up into short pieces called chromosomes. Each type of organism has its own specific number of chromosomes (and chromosomes are paired). Example: Humans have 23 pairs of chromosomes, therefore humans have a total of 46 chromosomes. Two of those chromosomes determine sex. The others carry information about other things like height, eye color, and so on. The process of cell division (mitosis or meiosis) can be broken down into several steps. Those steps are described below.

## MITOSIS

Step 1: INTERPHASE–DNA breaks up into short segments called CHROMOSOMES. Each chromosome "runs a copy" of itself. Each pair of copies stays attached in the middle only (the CENTROMERE). Extra cytoplasm is made.

Step 2: PROPHASE–Chromosomes thicken and shorten, the nuclear membrane disappears, and threadlike spindle fibers appear at the ends of the cell.

Step 3: METAPHASE–Chromosomes line up along the middle of the cell and each centromere is attached to a spindle fiber.

Step 4: ANAPHASE–Spindle fibers pull one chromosome from each pair towards opposite ends of the cell.

Step 5: TELOPHASE–The nuclear membrane reappears and the cell pinches inwards in the middle (FURROWING in animal cells or DIVISION PLATE FORMATION in plant cells). Two DAUGHTER CELLS are produced.

PROCEDURE: MITOSIS (Produces all body cells except egg and sperm cells.)

    1. Compare the descriptions of mitosis above to the diagrams below. Label each diagram as a step of mitosis.

 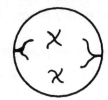 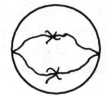 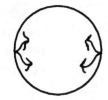 

_____    _____    _____    _____    _____

35

Date: _____ Names: _____

2. Use the diagrams on the previous page as a guide to construct a model of mitosis using yarn or short pieces of pipecleaner as chromosomes.

3. Get the following items (if they are not at your lab station):
   a. a section of yarn (or pipecleaner)
   b. a spool of thread
   c. a bottle of glue
   d. a marker
   e. a compass

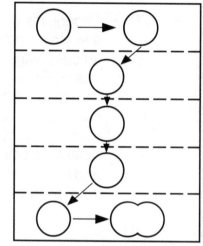

4. On your construction paper:
   a. Divide the paper into five sections.
   b. Draw large circles (cells) in each of the sections. (See diagram at right.)

5. Follow the instructions below to add or move pieces of yarn before gluing them into place on your construction paper.

Interphase–(first circle) Glue two pieces of yarn inside the cell. Draw the nuclear membrane.
(second circle) To show that the chromosomes duplicate, add two chromosomes (pieces of yarn tied together in the center) to the cell. Glue them into place. Draw the nuclear membrane.

Prophase–Position the chromosomes just like you did in the previous phase. Add spindle fibers (thread) attached from the ends of the cell to the centromeres of the chromosomes (where the two sections intersect). Glue them into place.

Metaphase–Position the chromosomes along the middle of the cell with spindle fibers attached. Glue them into place.

Anaphase–Place the separated chromosome copies from each pair (still attached to spindle fibers) on opposite sides of the cell. Glue them into place.

Telophase–(first circle) Position the chromosomes without spindle fibers on two sides of the cell. Glue them into place. Draw a dotted line showing furrowing of the cell.
(second circle) Draw in the nuclear membranes. Position the chromosomes in the new daughter cells. Glue them into place.

***LABEL ALL PHASES OF MITOSIS THAT YOU HAVE CONSTRUCTED. TURN IN YOUR POSTER.***

Date: _____ Names: _____

<u>MEIOSIS</u> (Produces only egg and sperm cells)

1. Cells undergoing meiosis go through the same first steps described in mitosis, except that process is called MEIOSIS I, not mitosis.

2. If cells are to become egg or sperm cells, after meiosis I, when there are two daughter cells, those daughter cells then undergo meiosis AGAIN. The only difference is that the chromosomes are not copied in the interphase stage.

3. All other stages of this second meiosis are the same. This series of divisions is called MEIOSIS II.

4. The newly formed cells end up with half the normal number of chromosomes. If not, during fertilization when the egg and sperm cells join, there will be twice as many chromosomes as the organism needs, and it will not survive.

5. In the spaces below, draw the steps involved in MEIOSIS II. Assume that meiosis I (which is just like mitosis) has already occurred.

QUESTIONS

1. How many chromosomes do humans have?_____

2. List the stages of mitosis in order from first to last?

a. _____          b. _____

c. _____          d. _____

e. _____

3. How does understanding mitosis help you understand meiosis?_____
_____

4. How is the end product of meiosis different from that of mitosis?_____
_____

5. Why is it that meiosis (also called "reduction division") must occur to form egg and sperm cells?
_____
_____

Date: _____ Names: _____

# OBSERVING CELL STRUCTURES

INTRODUCTION: What is inside a tiny blob-like cell that keeps it alive?

OBJECTIVE: The cell is the basic unit of life. Anton van Leeuwenhoek (in the 1700s) developed lenses that could magnify cells. Since then much has been learned about cell structure. Cells contain even smaller structures that work together to keep them alive. These structures are called ORGANELLES, because they work like the organs in our bodies to keep us alive. In this activity, we will attempt to locate several of the more prominent organelles.

PROCEDURE: <u>CHEEK CELLS</u>

1. Wipe and clean a microscope slide.

2. Place one drop of water in the center of the slide.

3. Gently scrape the inside of your cheek with a toothpick to remove some cells from the inner lining of your mouth.

4. Rinse the cheek cells off into the water droplet.

5. Add one drop of iodine to the solution.

6. Cover it with a coverslip—let the coverslip fall over onto the droplet.

7. Focus the cells on the highest power possible, and sketch them in the circle below.

8. Label the dark-stained NUCLEUS, the CELL MEMBRANE, and the CYTO-PLASM.

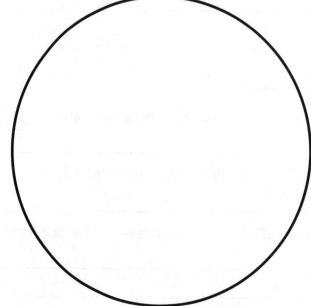

38

Date: _____ Names: _____

## ONION CELLS

1. Get an onion section from those provided.

2. From the bottom side of the onion section peel off a piece of the thin "skin"—about one centimeter (cm) square.

3. Place this skin in the middle of a clean microscope slide.

4. Add one drop of iodine to the skin.

5. Cover the droplet with a coverslip—let the coverslip fall onto the droplet.

6. Focus the cells on the highest power possible, and sketch them in the circle below.

7. Label the NUCLEUS, CELL WALL (MEMBRANE), CYTOPLASM, and VACU-OLES.

Date: _____ Names: _____

QUESTIONS:

1. Why is iodine used in these slide preparations?_____

_____

_____

2. List as many different cell structures as you were able to see in the slides.

_____     _____     _____

_____     _____

3. Why are these cell structures called "organelles"? _____

_____

_____

Date: _____ Names: _____

# OSMOSIS AND DIFFUSION

INTRODUCTION: If you break a bottle of perfume, eventually the scent will spread throughout the whole room, right? If you hang around school long enough eventually you will "absorb" some knowledge, right?

OBJECTIVE: Chemicals move through cells by two processes called OSMOSIS and DIFFUSION. Materials move through the cell membrane by osmosis and then spread to areas of lower concentration by diffusion. In this activity, we will use beets to observe the effects of certain chemicals as they enter beet cells by osmosis and then spread by diffusion within the cells.

PROCEDURE:

1. Label three test tubes and set them up as follows:
    A: 5 mL acid
    B: 5 mL base
    C: 5 mL water

2. Get three small chunks of beet, and record their color in the chart below.

3. Place one chunk of beet in each test tube.

4. Record any immediate observations in the chart below.

5. Set the tubes aside for 15–20 minutes.

6. After 15–20 minutes, check the color of the beets in the test tubes, and record any observations in the chart below.

| TUBE | STARTING BEET COLOR | SET-UP BEET COLOR | FINAL BEET COLOR |
|------|---------------------|-------------------|------------------|
| A    |                     |                   |                  |
| B    |                     |                   |                  |
| C    |                     |                   |                  |

Date: _____ Names: _____

QUESTIONS

1. Why did we need to set up tube C? _____

_____

_____

2. Describe the effects of each chemical on the appearance of the beet cubes.

Acid:_____

_____

Base:_____

_____

Water: _____

_____

3. What evidence is there that the acid or base entered the beet cells by osmosis?

_____

_____

4. Why were beet cells good to use for this activity? _____

_____

_____

5. When perfume scent spreads to cover a whole room, what is that process called?

_____

6. "Absorbing" knowledge from being at school is based on what scientific concept?

_____

7. Describe osmosis:_____

_____

_____

8. Describe diffusion:_____

_____

Date: _____ Names:_____

# PERCENT WATER CONTENT IN FOOD

INTRODUCTION: What happens to bread when you leave the wrapper open overnight? What happens to potato or cucumber slices left laying out?

OBJECTIVE: In this activity, we will calculate the percent of water content in several different types of food by measuring the amount of water lost due to evaporation.

PROCEDURE:

1. From the designated area, get one sample of each food listed in the chart.

2. Carefully weigh each food (to the nearest tenth of a gram), and record its weight in the chart under "Day 1 Weight."

3. Arrange the weighed food samples on a piece of foil or an aluminum pan that is labelled with your group's names and class period.

4. Place this tray underneath the heating lights overnight (or in a hot oven for 20 to 30 minutes if available).

5. At the end of the drying period, reweigh each food sample. Weigh again to the nearest tenth of a gram.

6. Record this weight under "Day 2 Weight."

7. Also record the amount of weight lost as water under "Weight Lost." This can be determined by subtracting the smaller "Day 2 Weight" from the larger "Day 1 Weight."

8. To calculate the percent of the food that was lost as water that evaporated, follow the example below.

   Weight Lost  ÷  Day 1 Weight   X  100 = Percent Water

   a. Divide "Weight Lost" by the "Day 1 Weight."
   b. Multiply this answer by 100.
   c. This result is the percent of water that made up that particular sample.

9. Record the final percent of water in the chart under "Percent Water."

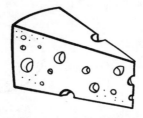

Date: _____ Names: _____

| FOOD SAMPLE | DAY 1 WEIGHT | DAY 2 WEIGHT | WEIGHT LOSS | PERCENT WATER |
|---|---|---|---|---|
| Potato | | | | |
| Bread | | | | |
| Cucumber | | | | |
| Pinto Beans (5) | | | | |
| Cheese | | | | |
| Meat | | | | |

QUESTIONS

1. Which food(s) was highest in water content? _____

_____

2. Which food(s) was lowest in water content? _____

_____

3. If food didn't contain so much water, what would you have to do to make up for it?

_____

_____

4. What happens to cause the bread to harden and the vegetables to shrivel?

_____

_____

_____

Date: _____     Names: _____

# WHY DO LEAVES CHANGE COLORS IN THE FALL?

INTRODUCTION: Why do some leaves turn yellow in the fall while some turn red or orange?

OBJECTIVE: Besides the green pigment, chlorophyll, there are other pigments in leaves that give them their respective colors. In the fall when the days become shorter and the temperatures become lower, the green chlorophyll pigment is no longer produced and breaks down. Then the other pigments, called carotenoids because of their orange, yellow, and red colors, become visible. In this activity, we will separate the pigments through a process called chromatography. Through this method, a solution containing leaf pigments from leaves ground up in alcohol is allowed to travel up a piece of filter paper. The smaller pigments travel higher. The different pigments can be identified by their colors.

PROCEDURE:

1. Trim the filter paper provided so that it has the dimensions of the paper in the diagram at the right. It should be long enough to stick out the top of the test tube (approx. 20 cm).
2. Cut a small notch on either side of the bottom end of the filter paper as shown in the diagram at the right.
3. On one sheet of filter paper, place a dot just above the notch in the middle of the paper using GREEN LEAF EXTRACT. (Make several more dots on top of the first dot to make sure there is a lot of extract there).
4. Blow the dot dry for about one minute.
5. Repeat steps 3 and 4 for the COLORED EXTRACT on the other filter paper.
6. Pour 5 mL of ACETONE into each of your two test tubes.
7. Hang one piece of filter paper inside each test tube so that only the end of the filter paper below the notch is in the acetone.
8. Place the test tubes in the test tube rack for about 15 minutes and observe the reaction. Record observations every three minutes in the chart below.

12 mm

18 cm

2 cm

| EXTRACT | 3 MIN. | 6 MIN. | 9 MIN. | 12 MIN. | 15 MIN. |
|---------|--------|--------|--------|---------|---------|
| GREEN   |        |        |        |         |         |
| COLORED |        |        |        |         |         |

Date: _____ Names: _____

       9. At the end of the 15 minutes, remove the pieces of filter paper from the test tubes of acetone.

    10. Tape each piece of filter paper onto the back of this sheet and label each of the following:

       A. Which extract was applied at the start.

       B. What individual colors can be identified.

## QUESTIONS

1. From your observations, what job do you think acetone did during this reaction?

_____

_____

2. What colors appeared from the separation of the green extract? _____

_____

_____

3. What colors appeared from the separation of the colored extract? _____

_____

_____

4. How does this comparison prove that the absence of chlorophyll allows other colors to appear in the fall?

_____

_____

5. What do you think causes some colored pigments to travel higher up the filter paper than others?

_____

_____

6. Why did you have to be careful not to let your ink spot touch the acetone? _____

_____

_____

7. Why do leaves change to the colors of yellow, red, and orange in the fall? _____

_____

_____

Date: _____ Names: _____

# BLOOD TYPING

INTRODUCTION: What blood type do you have? How do doctors know which blood types can be given to which people in the hospital?

OBJECTIVE: In this activity we will simulate blood transfusions in order to gain an understanding of which blood types may be received from/donated to which other blood types.

PROCEDURE:

1. Set up five cups as follows:

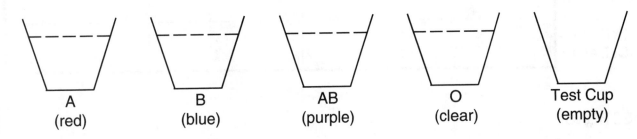

| A | B | AB | O | Test Cup |
| (red) | (blue) | (purple) | (clear) | (empty) |

2. Four of the cups should be filled with 3/4 cup of water of the designated color.

3. In order to determine which blood types can safely donate to/receive from which types, think of the different blood types as different colors. If a color can receive another color without a color change it is a SAFE transfusion. If not, it is UNSAFE.

4. To observe such reactions, place a small amount of one color of water in the empty test cup.

5. Carefully add another color to it. Remember that the cup that is pouring is "donating."

6. Shades may change during safe transfusions (example: clear "O" poured into red "A" will lighten the shade, but the color remains the same). Only complete color changes are unsafe (example: purple "AB" poured into red "A" causes a color change to purple and is unsafe).

7. Record your results in the chart.

8. Rinse and dry the cup that received the "transfusion" (the test cup).

9. Continue performing the transfusions until all of the combinations that are listed in the chart have been tried.

Date: _____ Names: _____

## RECEIVING

| | TYPE A (RED) | TYPE B (BLUE) | TYPE AB (PURPLE) | TYPE O (CLEAR) |
|---|---|---|---|---|
| **D O N A T I N G** TYPE A (RED) | | | | |
| TYPE B (BLUE) | | | | |
| TYPE AB (PURPLE) | | | | |
| TYPE O (CLEAR) | | | | |

QUESTIONS:

1. What blood types can TYPE A blood:

    a. receive? _____

    b. donate to? _____

2. What blood types can TYPE B blood:

    a. receive? _____

    b. donate to? _____

3. What blood types can TYPE AB blood:

    a. receive? _____

    b. donate to? _____

4. What blood types can TYPE O blood:

    a. receive? _____

    b. donate to? _____

5. Which blood type can give blood to ALL other blood types and is therefore called the "UNIVERSAL DONOR"? _____

6. Which blood type can receive blood from ALL other blood types and is therefore called the "UNIVERSAL RECIPIENT"? _____

Date: _____    Names: _____

# STARCH DIGESTION

INTRODUCTION: What happens to a Shredded Wheat biscuit once you take a bite and start chewing it? What does the biscuit turn into?

OBJECTIVE: In order to demonstrate the digestion of starch into sugar by saliva, we will test chewed and unchewed biscuits for the presence of starch and sugar.

** Iodine changes from a rusty brown color to dark purple/blue in the presence of starch.
** Benedict's solution turns yellowish-orange when heated in the presence of sugar.

PROCEDURE:
1. Label four test tubes as follows:
   A: unchewed—starch     B: unchewed—sugar
   C: chewed—starch       D: chewed—sugar
2. Crumble a Shredded Wheat biscuit and place 1/4 into test tube A and 1/4 into test tube B.
3. Add 3 mL of water to test tube A and 3 mL of water to test tube B.
4. Chew the remaining half biscuit and spit half into test tube C and half into test tube D.
5. To conduct the STARCH TESTS:
   a. Place three drops of iodine into test tube A and three drops into test tube C.
   b. Roll tube between hands to stir.
   c. Observe color changes.
6. To conduct the SUGAR TESTS:
   a. Add 1 mL Benedict's solution to test tube B and 1 mL to test tube D.
   b. Roll tube between hands to stir.
   c. Heat tube for 1–5 minutes. Make sure that the tube is tilted away from your lab partners and other people in the room.
   d. If one tube changes colors, you may stop heating both.
   e. Record color changes below.

| TEST TUBE | BISCUIT CONDITION | TEST PERFORMED | OBSERVATIONS/ RESULTS |
|---|---|---|---|
| A | Unchewed | Iodine test for starch | |
| B | Unchewed | Benedict's test for sugar | |
| C | Chewed | Iodine test for starch | |
| D | Chewed | Benedict's test for sugar | |

Date: _____ Names: _____

QUESTIONS:

1. In which conditions did the biscuit contain starch? _____

_____

_____

2. In which conditions did the biscuit contain sugar? _____

_____

_____

3. Describe positive test results for:

    a. starch. _____

_____

    b. sugar. _____

_____

4. What is the purpose in chewing half the biscuit? _____

_____

_____

_____

5. What happens to a biscuit once you start chewing it? _____

_____

_____

_____

Date: _____ Names: _____

# PROTEIN DIGESTION

INTRODUCTION: How does your stomach break down food when it digests it? What changes food in your stomach into tiny pieces that can be carried to cells to be used for energy?

OBJECTIVE: In this activity, we will investigate the digestion of a protein sample. Protein digestion usually begins in the stomach and is completed in the small intestine, where most digestion takes place. We will test several chemicals that are normally present in the human body in order to determine the best combination for digestion of protein.

PROCEDURE:
1. Label and set up four test tubes as follows:
   A: 10 mL water
   B: 10 mL pepsin
   C: 10 mL hydrochloric acid
   D: 10 mL pepsin and 2 drops hydrochloric acid
2. Place a small chunk of boiled egg white in each test tube. This is the protein sample.
3. Place a rubber band around all four test tubes and label them with your lab group's name.
4. Place the group of test tubes in the water bath underneath the heat lamps, and leave them overnight.
5. On the second day, observe the edges of the egg white for digestion, and record results in the chart below.

| TEST TUBE | CHEMICALS PRESENT | APPEARANCE OF EGG | AMOUNT OF DIGESTION |
|---|---|---|---|
| A | Water | | |
| B | Pepsin | | |
| C | Hydrochloric Acid | | |
| D | Pepsin and HCl | | |

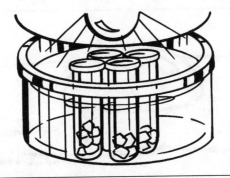

Date: _____ Names: _____

QUESTIONS:

1. What is the reason for using water in one of the test tubes? _____

_____

_____

2. In which test tube did the most digestion take place? _____

_____

3. What chemicals were present in that test tube? _____

_____

4. Why was it necessary to place the test tubes in a warm water bath that was above room temperature?

_____

_____

_____

5. Why did you have to wait overnight for this reaction to take place? _____

_____

_____

_____

6. What chemicals are likely present in the body for protein digestion? _____

_____

_____

Date: _____ Names:_____

# CALCULATING YOUR HORSEPOWER

INTRODUCTION: How powerful are you? How does your horsepower compare to that of a small car (about 100 HP)? What influences horsepower most: size or speed?

OBJECTIVE: In this activity, we will calculate the horsepower that you generate as you climb stairs.

PROCEDURE:

1. Measure the height of the stairs/steps that you will be running up in meters.

   _____ m

2. Have someone time you as you run up those steps for three trials. If possible, time to the nearest one hundredth of a second.

3. Total and calculate the average time by dividing by three.

   Trial 1 _____ seconds

   Trial 2 _____ seconds

   + Trial 3 _____ seconds

   _ _ _ _ _ _ _ _ _ _ _ _ _ _

   Total _____ seconds ÷ 3 = Average _____ seconds

4. Find your mass (weight) in kilograms by taking your actual number of pounds and dividing it by 2.2.

   _____ lbs. divided by 2.2 = _____ kg

5. Multiply your mass times the height of the steps.

   _____ kg X _____ m = _____

6. Divide the answer in #5 by your average time going up the steps from #3.

   _____ divided by _____ sec. = _____

7. Multiply this answer by 9.8 m/sec/sec to find how many Watts of power you produced.

   _____ X 9.8 = _____ Watts

Date: _____  Names: _____

8. Since there are 746 Watts in 1 Horsepower, divide your number of Watts by 746 to find your Horsepower.

_____ Watts divided by 746 = _____ HP!

QUESTIONS:

1. How does your horsepower compare to that of a small car? _____

_____

_____

2. If all of your classmates have the same horsepower as you do, how many students would it take to match the horsepower of a small car?

_____

Date: _____ Names: _____

# MICROBE CONTROL

INTRODUCTION: If you want to kill germs most effectively, which substance should you use for gargling or for cleaning?

OBJECTIVE: In this activity, we will investigate if mouthwashes and antiseptic cleansers actually kill "germs" as they advertise. Our "germs" will be yeast, and we will test several "germ killers." The indicator that will show us how many "germs" survived will be Methylene Blue. Methylene Blue changes color from dark blue to clear in the absence of oxygen. As yeast cells die, no oxygen is produced, and the solution will change color.

PROCEDURE:

1. Label seven test tubes as follows and add 2 drops of Scope, Listerine, alcohol, hydrogen peroxide, and chlorine to the appropriately labeled test tubes.

   C: Control            A: Alcohol
   C(ht): Heated Control  HP: Hydrogen Peroxide
   S: Scope              CHL: Chlorine
   L: Listerine

2. Place 20 drops of yeast solution in each test tube.
3. Heat the C(ht) test tube to boiling by placing it in a flame with the opening tilted away from people.
4. Add 2 drops of Methylene Blue to each test tube and roll each tube between your hands to stir it.
5. Put stoppers on the test tubes (or cover with Parafilm) so that they are airtight.
6. Place the test tubes in a designated location so that observations can be made over the next 10 class days.
7. Record color change observations in the chart provided.
8. If you have time, the experiment can be repeated with 5 drops each of Scope, Listerine, alcohol, hydrogen peroxide, and chlorine.

## CHANGES IN APPEARANCE

| Day | Control | Heated Ctrl. | Scope | Listerine | Alcohol | Peroxide | Chlorine |
|-----|---------|--------------|-------|-----------|---------|----------|----------|
| 1 | | | | | | | |
| 2 | | | | | | | |
| 3 | | | | | | | |
| 4 | | | | | | | |
| 5 | | | | | | | |
| 6 | | | | | | | |
| 7 | | | | | | | |
| 8 | | | | | | | |
| 9 | | | | | | | |
| 10 | | | | | | | |

Date: _____ Names: _____

QUESTIONS:

1. In the space below, rank the different solutions from darkest blue (most oxygen = most live yeast "germs") to lightest blue (least oxygen = fewest live yeast "germs").

| | | | | | | |
|---|---|---|---|---|---|---|
| | | | | | | |

DARKEST                                                                    LIGHTEST

2. What was the purpose of boiling one of the control solutions? _____

_____

_____

3. Under what two conditions were there the most living yeast "germs"?

_____

_____

4. Under what two conditions were there the least living yeast "germs"?

_____

_____

5. To kill germs most effectively, which solution should be used? _____

_____

# PROBABILITY TESTING

Date: _____ Names: _____

INTRODUCTION: What is the actual chance of having a female baby? What is the actual chance that a recessive gene will be expressed or that parts of recessive genes will appear in a genetic cross?

OBJECTIVE: In this activity, we will test to see if the predicted probabilities of certain events hold true in "real life" situations.

PROCEDURE:    I. SEX DETERMINATION
1. Given two possible outcomes, assume a 50–50 chance of either outcome occurring.
2. Use a coin to represent the two possible outcomes, as it may land on either heads or tails when it is flipped.
3. Flip a coin 50 times, and record below each time that it turns up heads or tails by marking out the letter representing the down side of the coin.

| h t | h t | h t | h t | h t | h t | h t | h t | h t | h t |
|-----|-----|-----|-----|-----|-----|-----|-----|-----|-----|
| h t | h t | h t | h t | h t | h t | h t | h t | h t | h t |
| h t | h t | h t | h t | h t | h t | h t | h t | h t | h t |
| h t | h t | h t | h t | h t | h t | h t | h t | h t | h t |
| h t | h t | h t | h t | h t | h t | h t | h t | h t | h t |

QUESTIONS:

1. How many times did the coin turn up heads? _____   tails? _____

2. How does this compare to/contrast with the expected 50:50 ratio? _____

---

II. SIMPLE CROSSES
1. Label two pennies with the head side "A" and the tail side "a."
2. Flip the two pennies at the same time for 40 times; record the number of times the pennies turn up "AA," "Aa," or "aa" in the spaces below.

|   AA   | Aa or aA |   aa   |
|--------|----------|--------|
|        |          |        |

QUESTIONS:

1. Real hybrid crosses would give the following frequencies (out of four flips): "AA" = 1; "Aa" or "aA" = 2, and "aa" = 1. How do your frequencies compare?_____

Date: _____ Names: _____

### III. DIHYBRID CROSSES

1. Use the two pennies from above labelled "A" and "a" on either side.
2. Use two other pennies provided and label the head side of each "B" and the tail side "b."
3. Flip all four coins at once for a total of 64 flips.
4. Record in the chart below the number of times each combination of coins occurs.

| Combination | Number of Occurrences |
|---|---|
| AA or Aa and BB or Bb | |
| AA or Aa and bb | |
| aa and BB or Bb | |
| aa and bb | |

QUESTIONS:

1. Real dihybrid crosses result in a 9:3:3:1 frequency in the outcome of certain combinations. How close were your frequencies? Explain.

_____

_____

_____

_____

_____

_____

_____

_____

# BIOLOGY ANSWER KEYS

## MITOSIS/MEIOSIS (page 37)
1. 46 chromosomes or 23 pairs
2. interphase, prophase, metaphase, anaphase, telophase
3. Meiosis undergoes the same steps as mitosis except for the second interphase where genetic materials are duplicated.
4. Meiosis produces new cells that have half the number of chromosomes as the original cells.
5. Meiosis produces egg and sperm cells by reducing the number of chromosomes by one-half so that when fertilization occurs, the number of chromosomes will equal the correct number for that organisms' body cells.

## OBSERVING CELL STRUCTURES (page 40)
1. Iodine is used to stain cell structures so that they are visible.
2. Cell structures that are visible will be: the nucleus, cell membrane, cytoplasm, nuclear membrane, and possibly vacuoles.
3. The structures are called organelles because they act like organs in our bodies, each doing a little job that helps keep the body alive.

## OSMOSIS AND DIFFUSION (page 42)
1. Tube C was used as a control so that it could be used to compare with the other tubes.
2. Answers will vary, but acid will probably have more digestion/color loss.
3. There will be a color loss when acid/base enters cells.
4. Beet cells are good to use because they have a dark color that fades due to osmosis.
5. Diffusion.
6. Osmosis.
7. Answers will vary, but should say something like, "Osmosis involves the movement of a substance across or through a membrane."
8. Answers will vary, but should say something like, "Diffusion occurs when a substance spreads from an area of higher concentration to an area of lower concentration."

## PERCENT WATER CONTENT IN FOOD (page 44)
1. Cucumber, potato, and bread should be highest in water content.
2. Beans, meat, and cheese should be lowest in water content.
3. You would have to drink a lot more water.
4. Water evaporates from the food.

## WHY DO LEAVES CHANGE COLORS IN THE FALL? (page 46)
1. Acetone dissolved the extract and carried it up the filter paper.
2. A green line and a faint brownish line should be visible.
3. Only the brownish line will be visible.
4. Chlorophyll is not present in the colored extract because no green line is visible.
5. Smaller pigments are more easily carried higher up the filter paper.
6. If the extract touches the acetone it dissolves into it and floats in the test tube instead of moving up the filter paper.

7. Leaves change colors in the fall because the chlorophyll breaks down and allows the other pigments that were already present to become visible.

## BLOOD TYPING (page 48)
1a. A and O; b. A and AB
2a. B and O; b. B and AB
3a. A, B, AB, and O; AB
4a. O; b. A, B, AB, and O
5. O
6. AB

## STARCH DIGESTION (page 50)
1. Starch was present before the cracker was chewed.
2. Sugar was present after the cracker was chewed.
3a. Iodine will change from brown to purple.
  b. Benedict's solution will change from blue to orange-yellow.
4. The other half can be tested as a control to compare with the samples that were chewed.
5. The digestion of the cracker begins with the breaking down of starches into sugars.

## PROTEIN DIGESTION (page 52)
1. The water is the control that other tubes are compared to.
2. Answers may vary, but pepsin and HCl should cause the most digestion.
3. Answers may vary depending on #2, but the expected result should be pepsin and HCl.
4. The warm water bath most closely simulates the temperature of the stomach.
5. Digestion of food takes about six hours to complete, so we had to wait a while to allow time for the reaction to occur.
6. Pepsin and HCl

## CALCULATING HORSEPOWER (page 54)
1. Answers will depend upon the calculated horsepower of the student.
2. Answers will depend upon the calculated horsepower of the student.

## MICROBE CONTROL (page 56)
1. Answers will vary according to the results observed.
2. Boiling one tube should provide a sterile setup in which no growth occurs—a control.
3. Answers will vary according to the results observed.
4. Answers will vary.
5. Answers will vary.

## PROBABILITY TESTING (pages 57–58)
 I. 1. Answers will vary.
    2. The expected result should be very close to the 50:50 ratio.
 II. 1. Answers will vary, but should be very close to the expected 1:2:1 ratio.
III. 1. Answers will vary, but should be very close to the expected 9:3:3:1 ratio.

# ECOLOGY INDEX AND MATERIALS LIST

Date: _____ Names: _____

# NATURAL SELECTION

INTRODUCTION: Why do some lizards change to the color of their surroundings? Why do zebras have stripes? How could it possibly be helpful to be as ugly as an alligator? All of these characteristics are helpful in hiding organisms from predators or in allowing them to hide while they hunt for prey.

OBJECTIVE: In nature, organisms (plants or animals) survive only if they are able to satisfy their needs and avoid disease, accident, or predation. Surviving organisms are said to be "fit" for survival. These "fit" organisms are able to bear offspring that are also likely to survive simply because they inherit helpful characteristics from their parents. This process is called "natural selection" or "survival of the fittest." In this activity, we will simulate the selection process and observe its effects.

PROCEDURE:

1. There are 100 yellow/gray chips in the cup at your lab station.
2. Cover the cup and shake it vigorously.
3. Dump the chips on the poster board provided, and spread them without turning them over from the color on which they land. All the chips should be within easy reach.
4. One lab partner should close his or her eyes and pick up one chip at a time for one minute.
5. Another lab partner should count the number of gray versus yellow chips "picked up" (that don't survive)—record color picked up in the chart below at the end of one minute.
6. Repeat this procedure two more times, recording the data in the chart below. Total and average the data for each color.
7. After three trials with eyes closed, you should conduct three trials in the same manner with eyes OPEN.
8. Record the data for each of the three trials in the chart below. Total and average the data for each color.

| | NUMBER OF YELLOW | | | | | NUMBER OF GRAY | | | | |
|---|---|---|---|---|---|---|---|---|---|---|
| | Trial 1 | Trial 2 | Trial 3 | Total | Average | Trial 1 | Trial 2 | Trial 3 | Total | Average |
| Eyes Closed | | | | | | | | | | |
| Eyes Open | | | | | | | | | | |

Date: _____ Names: _____

QUESTIONS:

1. With eyes closed, which color has the greatest average number of chips picked up?

_____

2. With eyes open, which color has the greatest average number of chips picked up?

_____

3. What accounts for the differences you may have observed in your data above?

_____

_____

4. If the two colors represent characteristics of organisms, what happens to the organisms that are:

     a. "selected"? _____

     b. not "selected"? _____

5. Which part of the activity represents:

     a. random selection? Why? _____

_____

     b. natural selection? Why? _____

_____

6. What characteristic may have caused the natural selection to occur for the chips? Why?

_____

_____

7. Besides color, what are some other characteristics that might lead to the selection of one member of a species over another?

     a. _____ b. _____ c. _____

     d. _____ e. _____ f. _____

8. Why do lizards change colors or zebras have stripes? _____

_____

9. What color would you expect a chameleon to be in grass? _____ On a tree trunk? _____

Date: _____ Names: _____

# PICKY EATERS

INTRODUCTION: How do all the different types of birds find enough food to survive? What happens if the pecan tree in your yard doesn't produce enough pecans for the blue jay that lives nearby? How has your cat changed the way the blue jay hunts? These are examples of things that animals face in order to find enough food to survive.

OBJECTIVE: Every portion of our environment, no matter how small, is likely to support some form of life. The number of individuals a particular area can support over a long period of time is known as its CARRYING CAPACITY. Factors, like the availability of food, will limit the growth of populations of organisms. Predators that consume organisms also limit the growth of certain populations. Organisms that simply cannot compete with other organisms usually develop some other strategy for growth and survival, like hunting only at night when predators sleep, or changing colors to blend in with their environment. In this activity, we will simulate feeding sessions of toothpick birds, gather data, and observe adaptations for survival.

PROCEDURE: <u>PART I—Food availability can limit the survival of members of a species.</u>

1. You will be assigned to a group of toothpick birds—you look for toothpicks for "food."

2. In the marked area are many toothpicks. You will have a time limit to find as many toothpicks of your assigned color as you can.

3. At the end of the time limit, add the number of all the toothpicks of your assigned color found by each member of your group and record this total in the third column of the chart on page 65.

4. Complete the sections of the chart by calculating how many members of your group are able to survive on the number of toothpicks found if 10, 20, or 30 toothpicks per bird are needed for survival. Record your conclusion in the fourth column of the chart as "_____ survive, _____ die."

Date: _____ Names: _____

| Colorbirds | # of Toothpicks Needed Per Bird | # of Toothpicks Found | Effects on Bird Group How Many Survive/Die? |
|---|---|---|---|
| YELLOW | 10 | | |
| | 20 | | |
| | 30 | | |
| NATURAL | 10 | | |
| | 20 | | |
| | 30 | | |
| GREEN | 10 | | |
| | 20 | | |
| | 30 | | |
| BLUE | 10 | | |
| | 20 | | |
| | 30 | | |
| RED | 10 | | |
| | 20 | | |
| | 30 | | |

QUESTIONS:

1. Which group found the most toothpicks (food)? Why?_____

_____

2. Which group found the fewest toothpicks (food)? Why? _____

_____

3. If food availability was the only thing influencing organism survival, which group above would survive:

a. best? _____ b. worst? _____

4. Besides the lack of ability to find food, what are some other things that might cause organisms not to survive?

a. _____ b. _____ c. _____

### PART II—Predators can affect the survival of organisms.

1. In this phase of the activity, you will remain in your colorbird group.

2. You will search in two different ways—independently and in a group (holding onto one another). When you hunt together, you may pick up toothpicks only when you are "attached" to your group. If you need to separate, you may, but you may not pick up toothpicks while separated from your group.

3. A predator will enter the "feeding area" and must walk heel-to-toe to stalk/chase prey. Anyone touched by the predator is dead, must throw all toothpicks back into the feeding area, and step outside the feeding area.

Date: _____ Names: _____

4. At the end of each time limit for hunting individually and in groups, record the total number of toothpicks found by the survivors of your group in the chart below.

5. Calculate how many members of your group survive or die for each hunting method. **Twenty toothpicks per organism are needed for survival.** Record as "___survive, ___die." You may also need to record the number of group members who became prey.

| Colorbird | Hunting Method | # Toothpicks Found | Effects on Bird Group<br># That Survive, Die, are Prey |
|---|---|---|---|
| YELLOW | Independently | | |
| | Collectively | | |
| NATURAL | Independently | | |
| | Collectively | | |
| GREEN | Independently | | |
| | Collectively | | |
| BLUE | Independently | | |
| | Collectively | | |
| RED | Independently | | |
| | Collectively | | |

QUESTIONS:

1. Which hunting method was:

   a. safest? Why? _____

_____

   b. most productive? Why? _____

_____

2. In general, how did the predator affect the number of toothpicks gathered?

_____

_____

66

Date: _____ Names: _____

3. Describe some methods you could have used to protect yourself from the predator while still being able to find enough food to survive.

a. _____

b. _____

4. If you could not find enough of your particular food type, describe some collecting methods you could use to keep from starving.

a. _____

b. _____

5. What would a blue jay that couldn't find enough pecans in your yard have to do to eat?

_____

_____

_____

6. What things would a blue jay need to do differently if your cat was hunting it?

_____

_____

_____

7. How might hunting with a whole flock of birds affect the blue jay's success at find food?

_____

_____

_____

# BEAR-LY ENOUGH: TEACHER PREPARATIONS

1. Orange, blue, yellow, red, and green construction paper or poster board will be needed.

2. Cut the paper or poster board into 2″ X 2″ or 2″ X 3″ pieces. For a classroom of 30 students, make 30 cards of each color as follows:

Orange      (nuts—walnuts, acorns, pecans, hickory nuts)
mark 5 pieces N–20, mark 25 pieces N–10

Blue      (berries—blackberries, elderberries, raspberries)
mark 5 pieces B–20, mark 25 pieces B–10

Yellow      (insects—grub worms, larvae, ants, termites)
mark 5 pieces I–12, mark 25 pieces I–6

Red      (meat—mice, rodents, peccaries, beaver, muskrats, young deer)
mark 5 pieces M–8, mark 25 pieces M–4

Green      (plants—leaves, grasses, herbs)
mark 5 pieces P–20, mark 25 pieces P–10

3. The numbers written on each card represent the numbers of pounds of food provided by each represented food card.

4. Black bear typically eat about 80 pounds of food in a ten day period or about 8 pounds of food per day.

5. Keeping these figures in mind, make and distribute the appropriate number of food cards for your size group of students. There should be less than 80 pounds of food per student so that there is not actually enough food in the area for all the bears to survive.

6. A fairly large area is required to do this activity. About 50 square feet would be a sufficient area in which to scatter the pieces of paper.

Date: _____    Names: _____

# BEAR-LY ENOUGH

INTRODUCTION: Suppose your mom sent you shopping and you could only carry one item at a time on your bicycle. Would you be able to bring back all the ingredients for dinner by the end of the day? What if you had to pedal with only one leg, or you had to ride blindfolded?

OBJECTIVE: The ability of a given area to supply adequate food, water, and shelter for a given number of organisms indefinitely is known as the CARRYING CAPACITY of that environment. In this activity, we will simulate bear feeding and the effects of carrying capacity on a group of bears.

PROCEDURE:
1. Students will be given envelopes that will represent their "dens."
2. Bears typically go out and eat, return to the den, and then go out and eat the next time they are hungry without stockpiling food in the den. Our simulation of that process will be collecting a single piece of "food," taking it to the envelope/den outside the feeding area, and returning to collect the next piece of "food."
3. The food is represented by different colored pieces of paper. Since bears are omnivores, they eat other animals, plants, nuts, insects, and berries.
4. Bears gather food rather than chase it down (in most cases); therefore, bears must walk heel-to-toe in this activity to collect their food.
5. There will be three special bears in the group:
    a: a young bear with a broken leg (must hop on one leg to search for food).
    b. a young bear blinded by a porcupine (must search blindfolded).
    c. a parent bear who must find twice as much food as the others to support two small cubs.
6. A typical bear requires **8 pounds of food per day to survive.** We will simulate a 10-day period in which 80 pounds of food must be found. Letters and numbers on the pieces of paper will signify what kind of food it is and how many pounds it represents.
7. A signal to start gathering will be given. When the round is completed (all "food" is picked up), each bear should total the amount of food in each category as well as the overall total for each trial and record the totals in the chart below.

| TYPE OF FOOD | NUMBER OF POUNDS OF FOOD GATHERED | | |
| --- | --- | --- | --- |
| | TRIAL 1 | TRIAL 2 | TRIAL 3 |
| NUTS = ORANGE | | | |
| BERRIES = BLUE | | | |
| INSECTS = YELLOW | | | |
| MEAT = RED | | | |
| PLANTS = GREEN | | | |
| Total Food | | | |
| Did This Bear Survive? | | | |

Date: _____ Names: _____

QUESTIONS:

1. What happened to the three special bears most of the time (at least two of the three times)?

     a. Bear with broken leg? Why? _____

_____

     b. Blind bear? Why? _____

_____

     c. Parent bear and cubs (parent will eat even if there isn't enough to also feed cubs)? Why?

_____

2. Why might the parent bear eat and let the cubs starve?

_____

_____

3. What techniques were used by the bears to find enough food to survive?

_____

_____

4. Overall,    a. how many bears were searching for food? _____

              b. how many bears found enough food? _____

              c. how many bears starved? _____

5. Was the carrying capacity of this area great enough to support this population of bears? What evidence is there for your answer?

_____

_____

6. What if the store your mom sent you to had only a single jellybean in the whole store? Would that be enough to feed your family for dinner?

_____

7. How might your family deal with such a situation? _____

_____

Date: _____     Names: _____

# FOOD CHAIN GAME

INTRODUCTION: If you let your cat out into the yard, what would it look for to eat? The blue jays that have to fly away from your cat are looking for what kind of food? Where does the pecan tree get the energy to produce pecans? All these things are connected in a food chain.

OBJECTIVE: Food chains and food webs are complex interactions between organisms competing for food. Food chains are constructed by recording who consumes whom in a certain environment. The lowest members in the chain are the producers because they get their energy by converting sunlight, carbon dioxide, and water into sugar (not by eating other organisms). Primary consumers are just above producers because they consume producers, like the cattle that eat grass. Secondary consumers are the organisms that eat primary consumers, like the wolf that eats the cattle. In this activity, we will simulate a food web involving lots of different organisms and record its appearance.

PROCEDURE:

1. Each student will be assigned an organism that he or she must "become" by acting like that organism. Example: grass stands still, but squirrels scurry around looking for nuts.

2. Each organism must try to capture as much prey as possible while trying to avoid being captured. Capture means that person is held onto and must move with you as you continue your search for more prey. If you are then captured by someone else, you and your prey become connected to your predator.

3. If there is doubt about which organism consumes which, consult the instructor.

4. An organism that is not captured survives only if it captures its own prey or is grass. Any grass not consumed automatically survives.

5. At the end of the activity, you should remain "attached" to your predator and all of your prey until your position in the food chain can be recorded.

6. After the chart is completed, sketch the food web/chain in the space provided.

SAMPLE FOOR CHAIN:

hawk

rabbit                    coyote

grass

Date: _____     Names: _____

| ORGANISM | PREY CONSUMED | PRODUCER, PRIMARY, OR SECONDARY CONSUMER | PREDATOR |
|---|---|---|---|
| GRASS | | | |
| ACORN | | | |
| CORN | | | |
| GRAIN | | | |
| RABBIT | | | |
| LEOPARD | | | |
| FOX | | | |
| SNAKE | | | |
| BEAR | | | |
| REDBIRD | | | |
| WOLF | | | |
| SPARROW | | | |
| DEER | | | |
| BOBCAT | | | |
| LION | | | |
| COW | | | |
| MOUSE | | | |
| ANTELOPE | | | |
| SQUIRREL | | | |
| FLOWER | | | |
| HAWK | | | |
| EAGLE | | | |
| HORSE | | | |
| CRICKET | | | |
| MOSQUITO | | | |
| CHIPMUNK | | | |
| FROG | | | |
| CHEETAH | | | |

Date: _____ Names: _____

CLASS FOOD WEB/CHAIN:

Date: _____ Names: _____

QUESTIONS:

1. Which organisms were not consumed? Why?

    a. _____

    b. _____

    c. _____

    d. _____

2. Which organisms were most quickly consumed? Why?

    a. _____

    b. _____

    c. _____

    d. _____

3. Why are plants called the "basis of the food chain"? _____

_____

4. How does the size of a predator relate to the size of its prey? _____

_____

5. From this activity, name examples of the following:

    a. producers          _____  _____  _____

    b. primary consumers  _____  _____  _____

    c. secondary consumers _____  _____  _____

6. How do the cat, blue jay, and pecan fit into a food chain? _____

_____

7. What would happen to the blue jay population in your yard if there were no cats? _____

_____

Date: _____   Names: _____

# PREDATOR AND PREY

INTRODUCTION: When your cat stalks a bird, how does it do it? Does the cat just walk up and grab the bird? What if the bird was a part of a large flock that would run and attack your cat? How would your cat react? Such behaviors are examples of how animals in the wild catch prey or avoid capture by predators.

OBJECTIVE: In the environment, organisms must compete for the food that they need to survive. Usually, unless they are plants, this means that they eat one another. Some organisms become the prey (those that are eaten) and others become the predators (those that eat others). In this activity, we will simulate predator-prey interactions. Most students will act as prey and will be assigned a particular noise to make in order to find others of the same "species" to band together for protection. Four predators will be released into the designated area. Predators may select any prey to "consume" by walking heel-to-toe and tagging the prey. Prey may do whatever is necessary to avoid the predator as long as they remain in the designated area.

PROCEDURE:

1. The entire class should form a circle facing outward.

2. Noises or noise-making devices will be assigned and predators will be selected.

3. The prey will be released in the designated area where they will search for a minimum amount of food (**50 pipe cleaners**), as well as try to avoid the predators.

4. Prey (holding onto each other) in groups of at least four **will not be approached by predators.**

5. After a 20-second head start for the prey, the predators will be released. Each predator must capture four prey to survive.

6. Captured prey must surrender their noise-making devices (if issued) and step outside the feeding area.

7. At the end of the time limit, each surviving individual should report how he or she avoided the predator while finding enough food to survive, and each predator should identify prey captured, as well as how the individual was captured.

Date: _____ Names: _____

| PREDATORS | PREY CAPTURED | METHOD USED TO CAPTURE PREY |
|---|---|---|
| | | |
| | | |
| | | |
| | | |
| | | |
| | | |
| | | |
| | | |
| | | |
| | | |
| | | |
| | | |
| | | |
| | | |
| | | |
| | | |

SURVIVING PREY          METHODS USED TO AVOID CAPTURE BY PREDATORS

| | |
|---|---|
| | |
| | |
| | |
| | |
| | |
| | |
| | |
| | |
| | |

Date: _____ Names: _____

QUESTIONS:

1. Why do predators usually avoid large groups of prey? _____

_____

2. What are some methods used by predators to capture prey?

    a. _____

    b. _____

    c. _____

3. What are some methods used by prey to escape predators?

    a. _____

    b. _____

    c. _____

4. If a new organism is introduced into an area with no natural predators, what is likely to happen:

    a. to its population?_____

    b. if this effect continues for a long time? _____

5. What could be done to prevent the situation in 4a and 4b? _____

_____

6. What are some benefits of predators?

    a. _____

    b. _____

    c. _____

7. What would happen in your neighborhood if no cats hunted birds?_____

_____

_____

Date: _____ Names: _____

# DEADLY CONNECTIONS

INTRODUCTION: What happens when you eat seafood or chicken that has bacteria in it? You get food poisoning, right? Just imagine what would happen if you ate food that contained pesticides. What would it do to you?

OBJECTIVE: Pesticides are chemicals that have been developed to stop the growth, reproduction, or life of organisms that are considered to be harmful or annoying. When the pesticides involve the use of poisons, the poisons frequently end up going where they are not wanted (like from an orchard to a lake). Many pesticides (poisons) persist in food chains as one organism consumes another. In this activity, we will simulate a food chain consisting of grasshoppers, shrews, and hawks. In this simulation, we will observe the effects of the pesticide as it affects the different levels of this food chain.

PROCEDURE:

1. Students will be divided into groups of two hawks, six shrews, and 18 grasshoppers.

2. Each grasshopper will be given a bag that will represent his or her "stomach."

3. "Food" is represented by different colored pipe-cleaners scattered throughout the designated area.

4. Grasshoppers will have one minute to collect as much food as possible by placing it in their bags.

5. After one minute, the shrews will be released into the feeding area to hunt grasshoppers. If caught, the grasshoppers must give the shrew their bags because when the shrews "eat" the grasshoppers, they also "eat" what is in their stomachs.

6. After 15 seconds, the hawks will be released to hunt the shrews (but not the grasshoppers). Captured shrews must give any collected grasshopper food to the hawk that captured them.

7. Everyone should continue to search for food until they are captured or the time limit expires.

8. At the end of the round, everyone with food should count the number of white pipe cleaners gathered and then count the number of multicolored pipe cleaners gathered and report those.

9. Multicolored pipe cleaners represent food contaminated by pesticides.

    a. If any grasshopper survived, but had half or more of his "food" made up of multicolored pipe cleaners, he or she dies.

Date: _____   Names: _____

   b. If any shrew survived, but had half or more of his "food" made up of
      multicolored pipe cleaners, he or she dies.
   c. The one hawk with the highest number of multicolored pipe cleaners will not
      die now, but the eggs produced by it or its mate next season will have shells
      so soft that they will fall apart. The other hawk will not suffer any
      consequences at this time.

| SURVIVING ORGANISM | # WHITE "FOOD" | # MULTICOLORED "FOOD" | RESULTS |
|---|---|---|---|
| Hawk #1 | | | |
| Hawk #2 | | | |
| Shrew #1 | | | |
| Shrew #2 | | | |
| Shrew #3 | | | |
| Grasshopper #1 | | | |
| Grasshopper #2 | | | |
| Grasshopper #3 | | | |
| Grasshopper #4 | | | |

QUESTIONS:

1. Describe several reasons why people might use pesticides.

   a. _____

   b. _____

   c. _____

2. How did the shrews and hawks become affected by the pesticides when they did not actually
consume contaminated plants?

_____

Date: _____ Names: _____

3. Why does the hawk suffer less (soft egg shells instead of death) when the grasshoppers and shrews are likely killed by pesticides?

_____

_____

_____

4. Think of another, similar food chain that might begin with the pesticide contamination of a plant and would include three other organisms.

_____

_____

_____

5. Name some alternatives to using pesticides to get rid of unwanted organisms.

　　　a. _____

　　　b. _____

6. Describe how pesticides may be spread from one area where they are "needed" to other areas where they are not needed (like from a field of peas to a lake)?

　　　a. _____

　　　b. _____

　　　c. _____

7. How do humans avoid getting food poisoning? _____

_____

_____

8. How should humans avoid poisoning themselves with pesticides? _____

_____

_____

Date: _____ Names: _____

# EN-DEER-ING UPS AND DOWNS

**INTRODUCTION:** If you and your three best friends were stranded on an island that had one shade tree, a five-gallon puddle of fresh water, and an apple tree that produced three apples each day, what would happen to the four of you? What things might cause conflict among you?

**OBJECTIVE:** A variety of factors affect the ability of wildlife to successfully reproduce and to maintain their populations over time. These factors are called LIMITING FACTORS. Disease, predatory/prey relationships, varying impacts of weather conditions from season to season, accidents, environmental pollution, and habitat destruction are among these factors. In this activity, we will simulate the interaction of a population of deer with the limiting factors of food, water, and shelter.

**PROCEDURE:**

1. Students will be assigned a number 1, 2, 3, or 4. One of the numbers will be designated to represent deer in the opening round of play. The other numbers will represent either food, water, or shelter.

   Food is simulated by placing both hands across the stomach.
   Water is simulated by placing both hands at the throat.
   Shelter is simulated by placing both hands on top of or over the head.

2. Food, water, and shelter begin play at the left side of the designated area with the symbol of whatever they have been picked to represent. All of these limiting factors should face away from the line of deer.

3. Deer begin play at the right side of the designated area displaying the symbol of the one of the three limiting factors for which they are searching. Deer's backs should be toward the line of limiting factors.

4. On a signal, both lines turn to face each other. Limiting factors stand still, displaying their symbols. Deer run toward the limiting factors whose symbols match their own, grab them, and take them back to the deer line. Other limiting factors not chosen (used) by the deer remain in place for the next round. Any deer not finding what it needs "dies" and becomes part of the habitat—either food, water, or shelter—for the next round.

5. Record the new number of deer and members of the limiting factors groups for round one.

6. Repeat this game nine more times, and record the results in the chart on page 82.

7. Graph two separate lines—one representing the number of deer in each round and one representing the number of members in the limiting factors group—on the graph on page 82.

Date: _____ Names: _____

| ROUND NUMBER | # DEER IN POPULATION | # OF LIMITING FACTORS |
|:---:|:---:|:---:|
| 0 | | |
| 1 | | |
| 2 | | |
| 3 | | |
| 4 | | |
| 5 | | |
| 6 | | |
| 7 | | |
| 8 | | |
| 9 | | |
| 10 | | |

## POPULATION OVER TIME

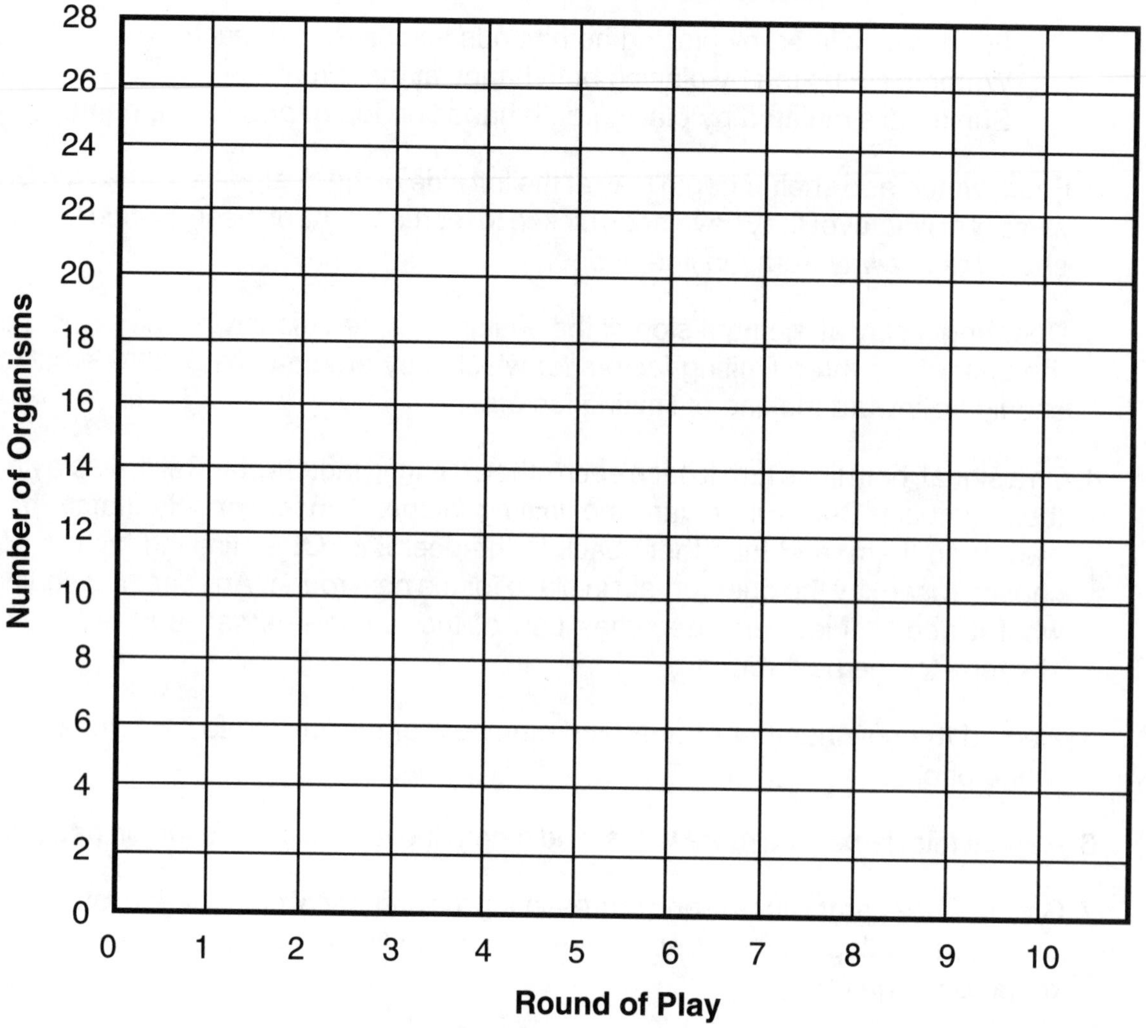

**Number of Organisms** (y-axis: 0, 2, 4, 6, 8, 10, 12, 14, 16, 18, 20, 22, 24, 26, 28)

**Round of Play** (x-axis: 0, 1, 2, 3, 4, 5, 6, 7, 8, 9, 10)

82

Date: _____ Names: _____

QUESTIONS:

1. What are some things that animals need to survive?

    a. _____    b. _____

    c. _____    d. _____

2. From the pattern of the lines on the graph, the populations of both the deer and the habitat (food, water, and shelter) tend to do what?

_____

3. How is the habitat affected when the deer population:

    a. increases? Why?_____

_____

    b. decreases? Why? _____

_____

4. What is happening to the deer population when the habitat is:

    a. increasing? Why? _____

_____

    b. decreasing? Why? _____

_____

5. What is a current method used to keep deer populations from growing so large that they use up their habitat (especially food) and begin to die?

_____

_____

6. If you could bring resources from elsewhere, how could you have changed the deserted island to make it a paradise where you and your friends could live indefinitely?

_____

_____

Date: _____ Names: _____

# NEIGHBORHOOD

INTRODUCTION: Suppose someone wanted to put a mall where the park near your house is. How would you feel about it? How would your science teacher feel about it? How would store owners feel about it?

OBJECTIVE: This activity is designed to provide students an opportunity to join a group that has a particular position regarding environmental issues. It also allows students an opportunity to discuss and debate those issues as they relate to a hypothetical community in which they live.

PROCEDURE:

    1. Groups include:

          Environmentalists

          Farmers

          Chamber of Commerce/Merchants' Association

          Parent-Teacher Association

          Industry

    2. The neighborhood consists of a farm and wooded area that are separated from a mill, the town, and a park area by a river.

    3. Issues include:

        a. Industry wants to build a shopping center where the park is located.

        b. Environmentalists want to prevent farmers from using pesticides on their crops.

        c. The Chamber of Commerce/Merchants' Association wants to allow the construction of a new road into town that would cut through the forest.

        d. The PTA wants to build an outdoor classroom in the forest.

        e. Industry wants to build a dam for hydroelectric energy just upstream from the mill, which is located outside of town.

        f. The farmers want to cut the forest and farm the area.

        g. The federal government has offered 10 million dollars to be allowed to place a toxic waste dump north of town.

        h. The state government wants to allow the construction of a housing development where the park is located.

        i. The Chamber of Commerce/Merchants' Association wants to build a mall where the park is located.

Date: _____     Names: _____

4. Students should be allowed to select which group they want to join (as long as numbers are fairly even and all groups are represented).

5. A map may be used as a visual aid.

6. Once the issue is presented, if the issue does not define the group's position, each group will have 30 seconds to decide and report their position.

7. At the end of two minutes, arguments should be heard. Each group should have one minute to orally list reasons to support their position. Groups should not be interrupted, and timing is critical.

8. After each group has listed the reasons for their positions, each group should have 30 seconds to respond to what the other groups have said by adding to their own list of reasons.

9. While reasons are being given, the instructor should record them in two columns, "For" and "Against." The list with the most DIFFERENT reasons determines whether the issue is passed or is defeated.

Date: _____    Names: _____

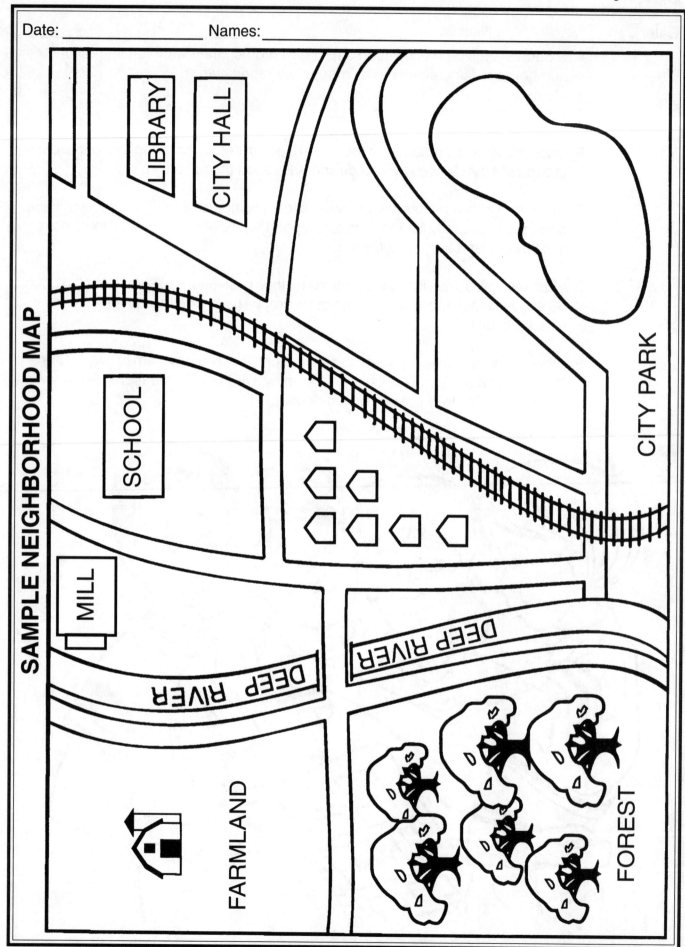

SAMPLE NEIGHBORHOOD MAP

LIBRARY

CITY HALL

SCHOOL

MILL

CITY PARK

DEEP RIVER

DEEP RIVER

FARMLAND

FOREST

Date: _____    Names: _____

# EXPLORING EARTH'S LAND-BASED BIOMES

**INTRODUCTION:** ECOLOGY is the study of living and non-living parts of the environment and how they affect each other. The environment can be divided into units, or parts, that have similar plants, animals, and climates. These units are called BIOMES. Biomes can be divided into HABITATS, which are the general areas where specific plant or animal species live. The specific area within the habitat where a particular species lives is called its NICHE.

**OBJECTIVE:** In this activity, students will research the land-based biomes listed and then match the plant or animal species with the biome in which it lives.

**PROCEDURE:**

1. Do some research about these land-based biomes. You may need to consult an encyclopedia or a book dealing with ecology and ecosystems. Information about the plant and animal species may also help.

2. On the line next to the plant or animal species place the letter of the appropriate biome from the word bank. Each biome will be used more than once.

| | | |
|---|---|---|
| **a. Tundra** | **b. Taiga** | **c. Temperate Forest** |
| **d. Tropical Rain Forest** | **e. Grasslands** | **f. Desert** |

_____ 1. Camel                          _____ 2. Raccoon

_____ 3. White Oak                    _____ 4. Purple Coneflower

_____ 5. Cooper's Hawk            _____ 6. Lichen

_____ 7. Bison                           _____ 8. Peccary

_____ 9. Prairie Dog                 _____ 10. Caribou

_____ 11. Wolverine                 _____ 12. Scorpion

_____ 13. Scarlet Macaw          _____ 14. Kapok Tree

_____ 15. Sugar Maple             _____ 16. Mesquite Tree

_____ 17. Musk Ox                    _____ 18. Pileated Woodpecker

_____ 19. Douglas Fir               _____ 20. Burrowing Owl

Date: _____ Names: _____

# ECOLOGY CROSSWORD PUZZLE

Use the clues below to complete the crossword puzzle on page 89. Students may need to do some research about ecology to find the information.

**ACROSS**

1. organisms that consume primary consumers (two words)

7. an organism that eats other organisms

9. organisms that consume producers (two words)

10. the specific area where a particular species lives

11. parts of the environment that have similar plants, animals, and climates

14. _____ _____ affect the abilities of wildlife to successfully reproduce

and maintain their populations over time.

**DOWN**

2. the study of living and nonliving parts of the environment and how they affect each other

3. the principle that organisms that possess helpful characteristics are more likely to survive and

bear offspring (two words)

4. complex interactions between organisms competing for food that may involve several food

chains (two words)

5. the general area where a specific plant or animal species lives

6. _____ get their energy by converting sunlight, carbon dioxide, and water into sugar.

7. organisms that are eaten by others

8. the number of individuals a particular area can support over a long period of time (two words)

12. the young of an organism

13. the transfer of energy from one organism to another as one organism consumes another (two

words)

Date: _____ Names: _____

# ECOLOGY CROSSWORD PUZZLE

Use the clues from page 88 to complete the puzzle.

Date: _____     Name: _____

# SEARCHING FOR ECOLOGY WORDS

Find the listed words in the puzzle below. Words may be printed vertically, horizontally, or diagonally and may read forwards or backwards. All of the words are associated with ecology in some way.

```
V W U C Y J O M T E C O L O G Y I F U Z
W S O S R E M U S N O C Y R A M I R P M
P P Q S E M O I B E O R G A N I S M H R
S O S E C O N D A R Y C O N S U M E R S
S U R V I V A L O F T H E F I T T E S T
E Q N O I T C E L E S L A R U T A N M U
N V E G A W I Y V O F O O D C H A I N J
H F P W Y K J G N V M G X U J U S X V F
W O N R F U S E H C I N V E W E A W R B
A O M W O A E R I M W S I D R N E M E C
G D Y Y N D B Q I K L V F V Y F D Y P M
X W C G F B U C Q B B B N M O J J L R G
D E A N R A I C K O Y R C V W R N T O L
S B W Q N E T U E K K P R E Y A E T D I
W X U D T Y N T F R X R V B J G Z S U T
J B H W G X T E O B S G M X G X R O C R
W Q X W D P R E D A T O R B O O B Z T N
R C A R R Y I N G C A P A C I T Y T I F
C O F F S P R I N G T A T I B A H T O G
L I M I T I N G F A C T O R S S L L N H
```

BIOMES
ENERGY
HABITAT
NICHE
ORGANISM
PRIMARY CONSUMERS
SECONDARY CONSUMERS

CARRYING CAPACITY
FOOD CHAIN
LIMITING FACTORS
OFFSPRING
PREDATOR
PRODUCERS
SURVIVAL OF THE FITTEST

ECOLOGY
FOOD WEB
NATURAL SELECTION
OMNIVORES
PREY
REPRODUCTION

# ECOLOGY ANSWER KEYS

## NATURAL SELECTION (page 63)

1. Answers may vary, but expected results are that the number of colored chips is very similar when eyes are closed.

2. Answers may vary, but expected results are that there will be more chips that contrast with the background selected with eyes open.

3. When eyes are closed, color had no influence, but when eyes are open, the contrasting color "stands out" against the background causing it to be seen and therefore picked more easily.

4a. These organisms survive, reproduce, and pass on their traits to their offspring.

  b. These offspring do not survive to reproduce.

5a. Random selection occurs when eyes are closed. Color cannot influence selection of the chips.

  b. Natural selection occurs when eyes are open. Color may influence which chips are selected.

6. Natural selection was caused by colors that contrast with the background. Contrasting colors are easier to see than chips that are the same color as the background.

7. Answers may vary but might include:

  a. speed;

  b. ability to fly;

  c. eyesight;

  d. intelligence;

  e. hunting skills;

  f. health.

8. They change color and have stripes to help them to blend in with their surroundings.

9. Green, gray/black

## PICKY EATERS

### PART I (page 65)

1. Answers may vary, but yellow will probably be found in greatest abundance. This happens because it contrasts with the green (or brown) grass and is easily seen.

2. Answers may vary but green and natural will probably be found least frequently. They blend in with either the live or dead grass and are difficult to see.

3a. Yellow (or whichever color was found in greatest abundance).

  b. Green or natural (or which ever color was found least).

4. Answers may vary but might include: a. predators; b. no nesting sites; c. hunting by humans.

### PART II (pages 66–67)

1a. Answer depends on which method resulted in fewest prey being caught.

  b. Answer depends on which method resulted in most food being collected.

2. In general, the predator should have decreased the number of toothpicks collected.

3. Answers may vary but might include: a. hunting when predator wasn't around, or b. outrunning the predator.

4. Answers may vary but might include: a. birds could search someplace else, or b. they could switch food types.

5. The blue jay would have to look someplace else or switch types of food.

6. The blue jay would have to watch out for the cat while hunting and hunt away from the cat.

7. A whole flock could watch out for one another and warn one another or distract the cat.

91

## BEAR-LY ENOUGH (page 70)

1. Answers will depend on group data but the following is expected:
   a. Broken-legged bear doesn't survive because he/she can't move as well.
   b. Blind bear doesn't survive because he/she can't see food and must rely on chance to find it.
   c. Parent bear might survive but can't find enough food for the cubs, too.
2. It will let the cubs starve thinking it can have more cubs next year but the cubs wouldn't survive later if the parent allowed itself to starve.
3. Answers may vary but might include:
   a. bears collect the "biggest" food first;
   b. bears hunt as fast as they can.
4. Answers depend on class data.
5. The carrying capacity is not great enough for all the bears if any bears do not survive.
6. No.
7. Your family would have to go hungry or shop someplace else.

## FOOD CHAIN GAME (page 74)

1a. Lion; b. Eagle; c. Cheetah; d. Leopard, etc. do not get consumed. These organisms are at the tops of their food chains.
2a. Grass; b. Flowers; c. Rabbit; d. Birds, etc. are most quickly consumed because they are near the bottoms of their food chains.
3. Plants change sunlight, water, and carbon dioxide into sugar.
4. Usually large predators hunt prey smaller than themselves.
5a. grass, flowers
   b. cow, deer, antelope
   c. wolf, lion, hawk
6. The blue jay eats the pecan, and the cat eats the blue jay.
7. No cats might lead to an increase in the blue jay population.

## PREDATOR AND PREY (page 77)

1. If prey can, they will fight together against the predator to survive.
2. Answers may vary, but might include:
   a. cornering the prey;
   b. ganging up on the prey;
   c. going after the slowest prey.
3. Answers may vary, but might include:
   a. hiding from the predator;
   b. ganging up to fight off the predator;
   c. outrunning the predator.
4a. Its population will increase.
   b. Disease and overuse of the habitat will eventually cause the organisms to start dying off.
5. Avoiding the introduction of organisms where there are no natural predators, or introducing a natural predator along with the organism will prevent the situation.
6. Answers may vary but might include:
   a. they regulate and stabilize prey populations;
   b. they consume weak, sick, or injured organisms;
   c. they encourage natural selection of stronger, faster, or "better" prey.
7. The bird population would grow out of hand.

## DEADLY CONNECTIONS (page 79–80)
1. Answers may vary, but might include:
  a. to increase crop production;
  b. to improve the quality of crops;
  c. to improve quality of living (after getting rid of mosquitoes or flies).
2. The organisms that they consumed had eaten the pesticides.
3. The hawks are larger and to them the dosage of pesticide is lower because of their size.
4. Answers may vary, but an example might be: "We put garden dust on beans that are eaten by grasshoppers that are eaten by birds that might be eaten by our cats."
5. Answers may vary, but might include:
  a. introduce a predator to consume the pest;
  b. use plants that are offensive to insects by planting with crops (example: marigolds planted in gardens).
6. Answers may vary but might include:
  a. water run-off;
  b. windblown;
  c. pesticide seeps into groundwater supply.
7. Food is carefully handled, kept cold, washed well, and cooked thoroughly.
8. Avoid pesticide use or use as few pesticides as possible.

## EN-DEER-ING UPS AND DOWNS (page 83)
1. Answers may vary but might include:
  a. food;
  b. water;
  c. shelter;
  d. oxygen.
2. The lines on the graph mirror one another. When one goes up the other goes down.
3a. Habitat decreases. More deer use more food, water, and shelter.
  b. Habitat increases. The habitat can build back up when fewer deer use it.
4a. Deer population is decreasing. The habitat is building back up because there are fewer deer using it.
  b. Deer population is increasing. The deer are depleting the habitat.
5. Hunting seasons.
6. Answers will vary, but you might expect: "Add more fruit trees, import water, add animals, and build shelter."

## EXPLORING EARTH'S LAND-BASED BIOMES (page 87)
|  |  |
|---|---|
| 1. f | 2. c |
| 3. c | 4. e |
| 5. b | 6. a |
| 7. e | 8. d |
| 9. e | 10. a |
| 11. b | 12. f |
| 13. d | 14. d |
| 15. c | 16. f |
| 17. a | 18. c |
| 19. b | 20. e |

## ECOLOGY CROSSWORD PUZZLE (pages 88–89)

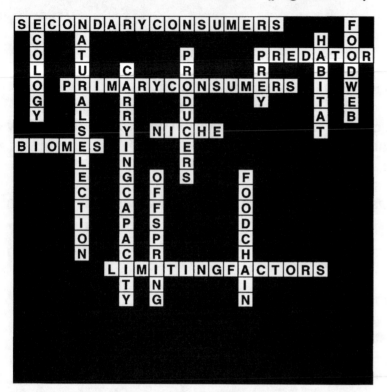

## SEARCHING FOR ECOLOGY WORDS (page 90)

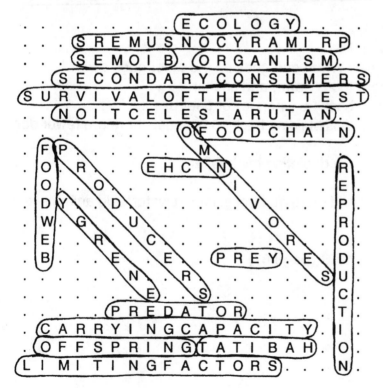